Industrial Flames

Vol. 1 Measurements in Flames

International Flame Research Foundation

INDUSTRIAL FLAMES

General editors:
Prof. J. M. Beér and Prof. M. W. Thring

Volume 1

Measurements in Flames

by J. Chedaille
and Y. Braud

Distributed in the United States by
CRANE, RUSSAK & COMPANY, INC.
52 Vanderbilt Avenue
New York, New York 10017

Edward Arnold, London
Crane, Russak, New York

© International Flame Research Foundation 1972

First published 1972
by Edward Arnold (Publishers) Ltd.

Published in the United States by:
Crane, Russak & Company, Inc.
52 Vanderbilt Avenue, New York, N.Y. 10017

Part 1 (Temperature measurement) is a
translation of IFRF Document Number
K20/a/33 (July 1966). Part 4 (Velocity
measurement) is a translation of IFRF
Document Number K20/a/34 (January
1967). Both have been translated from the
French by Dr N. A. Chigier of the
Department of Fuel Technology and
Chemical Engineering at the University of
Sheffield.

Library of Congress Catalog Card Number 72–80108

ISBN 0 8448 0012 0

Printed in Great Britain

Preface

'A very large fraction of the heat of combustion of solid, liquid and gaseous fuels is transferred by radiation and it would be of great value in saving fuel and improving appliance design if the laws of luminous radiation were as well understood as those of non-luminous radiation and convection. Luminous radiation is known to result from the process of combustion of hydrocarbons but the value of the resulting flame emissivity for any given thermal and mixing history of a mass of hydrocarbons is not known. Hence a research programme requires a study of the history of a mass of fuel under industrial conditions, and laboratory studies of the resulting behaviour of the soot together with the attempt to set up a mathematical theory. The first of these must be carried out on a furnace of nearly full scale and, since all possible flames cannot be investigated in complete detail, the investigations with such a furnace are divided into two kinds; engineering trials in which many flames are investigated with external measurements only and scientific trials in which a few flames are explored in detail.' (Quotation from summary of 'Research on flame radiation—the plan of an attempt to fill in an important gap in fuel technology,' by M. W. Thring published by *Fuel* xxix 1950, page 173.)

Attempts were made in 1938–9 by the BISF and the BCURA in Britain to establish an experimental furnace for the study of luminous radiation from producer gas in steelmaking furnaces and M. W. Thring was actually building one in London with controllable wall temperatures for studying P.F. flames, when World War II stopped all long range studies. M. Malcor in France was similarly concerned with flame heating in steel furnaces before World War II and during World War II Professor de Graaf made studies on radiation from open hearth flames using such simple apparatus as could be obtained. In 1947 de Graaf and Thring discussed their mutual interest at a meeting in Zurich and in 1948 de Graaf showed a furnace he had built for flames studies at the K.N.H.S. in IJmuiden to Thring. This led to a meeting in London at the end of 1948 between Malcor, de Graaf and Thring as a result of which it was agreed to support work at IJmuiden by setting up sponsoring committees in France, Holland and England.

A first trial with French and British workers cooperating with the Dutch furnace trial team carried out in 1949 was a study of radiation from air and steam atomised liquid fuel flames. This first cooperative trial may be considered to be the real birth of the International Flame Research Foundation as it was later called in its Deed of Establishment in 1955.

Since its inception, interest in the work of the Foundation has grown. At present it derives support from fourteen countries on four continents. Recognition of the work of the Foundation may be judged from the fact that the European Coal and Steel Community has provided generous financial assistance over the last ten years to accelerate the programme of research.

The function of the Foundation is stated in its Deed of Establishment as 'To obtain knowledge about and to gather experience of the combustion of gaseous, liquid and/or solid fuels, especially as far as this combustion aims at the heating of certain materials and to put this knowledge and this experience at the disposal of others for further development or industrial application.'

As a further step the scope of research has been extended to include studies of relationships between flame characteristics other than radiative properties, such as combustion length, flame shape, stability, combustion intensity, formation of pollutants and the occurrence of unburnt residual gases or solids on one hand and design and operational parameters (e.g. burner and furnace dimensions and fuel characteristics) on the other.

Studies have been made on pressure jet oil flames, pulverised coal flames of the types used in cement kilns and in boilers, natural gas flames, flames impinging on a cold surface and flames with swirl.

One of the early results of the work of the Foundation was the understanding of the role of the atomising fluid as a source of jet momentum which led to the possibility of burners of improved design giving the same jet momentum with a substantially reduced amount of such fluid.

Most of this work over the last 23 years has been published in scientific and technical journals, both in separate articles and also in special issues. The governing body of the Foundation has come to the conclusion that the work now constitutes a sufficiently comprehensive body of material that it would be of value to have it all available in organised form in a single book. The European Coal and Steel Community has given us every encouragement and support in this activity. Much of the work has required the development of special probes and other instruments suitable both for the IJmuiden type tunnel furnace and for use in practical boiler and furnace combustion chambers. It was therefore felt that the first volume should present all the information about these instruments in readily available form.

J. M. B.
M. W. T.

Contents

Part 1

Temperature Measurement

Introduction

In the study and control of combustion and heat transfer in industrial furnaces a knowledge of wall and gas temperatures is required. Heat transfer by radiation is dependent upon temperature to the fourth power and, under certain conditions, reaction rates are affected by temperature to the same power (1)(2). It is therefore important that measurements of temperature are made with a high degree of accuracy. Since conditions

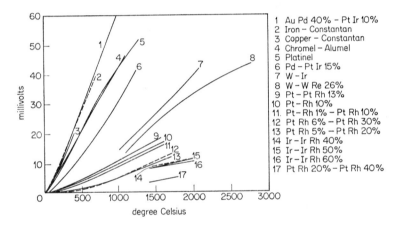

Fig. 1.1 Electrical properties of some thermocouples

under which measurements have to be made vary over a wide range it is often necessary to modify existing instruments in order to achieve high precision. It is for this reason that this chapter not only presents the instruments used at the Flame Research Foundation, but also includes the fundamental principles together with the causes and orders of magnitude of possible errors.

The measuring element in almost all these instruments is the thermocouple, which must be chosen, constructed and used with care as has been shown by a number of studies on the subject (3)(4)(5).

It is necessary to know the range of usage of each type of thermocouple (Fig. 1.1), as well as the errors that may result due to ageing of the thermocouple and due to negligence during operation, Fig. 1.2.

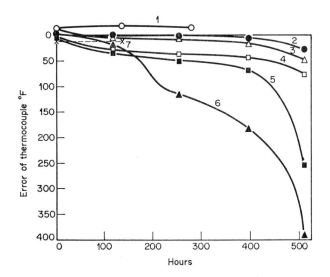

1. Pt Rh 6% − Pt Rh 30% 2 wires (0·25 mm) insulated with Al₂O₃
2.   ″   −   ″ ″ 1 wire Pt Rh 6% and a sheath Pt Rh 30% insulated with MgO
3.   Pt   − Pt Rh 10% 1 ″ Pt and a sheath Pt Rh 10% − ″   ″   MgO
4.   ″   −   ″ ″ 2 wires insulated with MgO in a complex sheath
5.   ″   −   ″ ″ ″ ″   ″   ″ ″ ″ ″ Pt Rh 10%
6.   ″   −   ″ ″ ″ ″   ″   ″ ″ ″ ″ ″
7.   ″   −   ″ ″ 2 ″   (1mm) insulated with Al₂O₃

Fig. 1.2 Comparison of performance of some thermocouples (diffusion and volatility effect)

Chapter 1

Surface Pyrometers

1.1 Surface temperature

When the surface is accessible, measurement can be made with surface pyrometers (4)(5):

(a) contact pyrometers, and
(b) optical pyrometers—total radiation, monochromatic disappearing filament, two colour, etc.

These instruments are simple to use and are suitable for discontinuous measurements.

1.2 Surface thermocouples

When the surface is not accessible, which is generally the case in industrial environments, or when it is desired to make continuous measurements, it is necessary to fix thermocouples in place before heating the surface. The method is simple, but care needs to be taken in order to avoid significant errors that can be introduced through the following:

1. Changes in local heat transfer conditions.
2. Uncertainty as to the position of the hot junction of the thermocouple.
3. Heat losses by conduction along the thermocouple wires when these are placed in strong temperature gradients.

All these errors are a function of the temperature gradients in the vicinity of the surface and the stronger these gradients are the more important it is to place the thermocouple carefully (4)(5).

Thermocouples should be selected having as small a diameter as possible, particularly when they are not subjected to any mechanical strain. Wire diameters of 0·1 mm are available and complete thermocouples with insulation and protective covering (Thermocoaxe) of external diameter 0·25 mm are made by Philips (Holland) and Pyro-electric

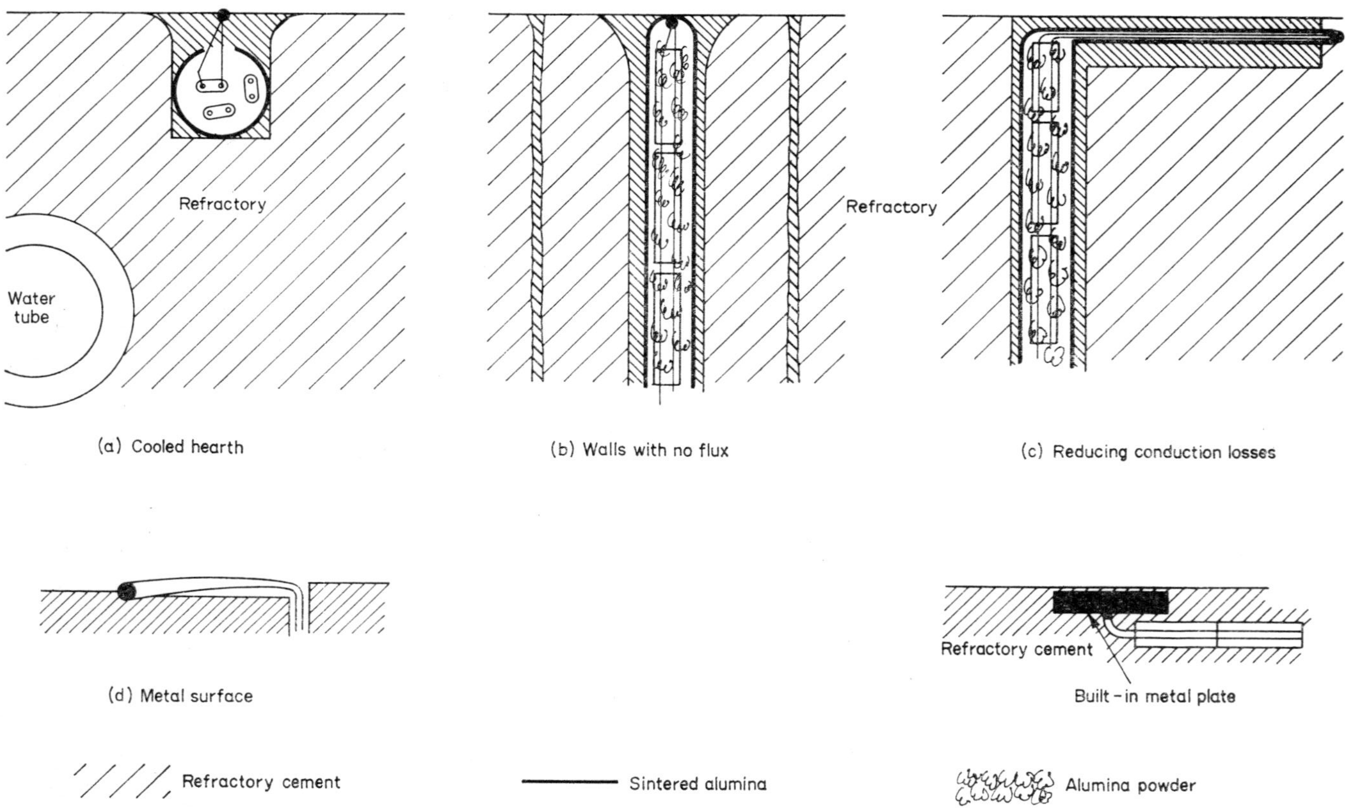

(a) Cooled hearth

(b) Walls with no flux

(c) Reducing conduction losses

(d) Metal surface

Water tube

Refractory

Refractory

Refractory

Refractory cement

Built-in metal plate

Sintered alumina

Alumina powder

Refractory cement

Fig. 1.3 Surface thermocouples

(U.S.A.). The error due to conduction along the thermocouple wires can be reduced by placing a certain length of wire near the hot junction, perpendicular to the temperature gradient (Fig. 1.3). For refractory walls the hot junction can be incorporated in an imbedded metal plate of high thermal conductivity and in which the temperature will not be appreciably affected by conduction losses. To obtain an accurate positioning of the measuring point for the case of a metal wall, the hot junction is often soldered onto the wall, or in a small cavity made in the wall, which can also serve as one of the branches in the thermoelectric circuit. For the particular case of heat flow in one direction only and when it is not required to measure the instantaneous temperature fluctuations at the surface, the surface temperature can be determined by extrapolation from two temperatures measured in the wall at known distances from the surface. Account must be taken of the conductivity of the wall material as a function of temperature.

1.3 Recording of temperatures

There are a large number of types of instrument capable of measuring with precision a difference in potential of a few millivolts. In practice, there are two that are of particular interest; the recording potentiometer and the datalogger.

Recording potentiometer

This instrument is perfectly suitable when the number of measurements to be made is small or when it is used for continuous control. It can be used with an automatic compensation for the cold junction and can make recordings from a number of thermocouples. Multipoint recorders are commercially available for 12 to 16 measurements, with the cycle of measurements being completed in less than 30 seconds.

Datalogger

Dataloggers are comparatively expensive but they can be recommended when the number of temperatures to be measured is large and they have subsequently to be introduced into calculations (6). The principle of this apparatus, which may consist of several hundreds of measuring points, is shown schematically in Fig. 1.4. Potential differences, generated by the thermocouples and automatically corrected for the cold junction, are transformed into degrees Celsius by a series of linear programmes. For the datalogger in use at the Foundation, the range is from $-5°C$ to $1600°C$.

   The calibration curve of the thermocouple is divided into a succession
of 12 linear elements. Measurement is made by a numerical volt meter and
the signal is transmitted to a printing or perforating machine. The system
can record, with a precision of 2°C, one temperature per second and the

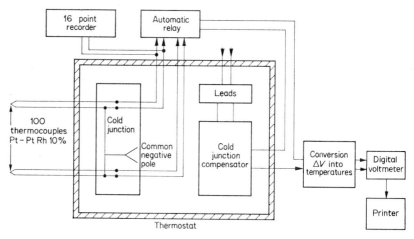

Fig. 1.4 Scheme of Solartron datalogger for recording wall temperatures

cycle of measurements can be repeated automatically, either continuously,
or at regular intervals (10 mins, 1 hour, 10 hours, etc.). When the results
are given on perforated tape, the tape can be introduced directly into a
computer. Human errors are completely eliminated without losing the
benefit of control, since the numerical volt meter displays the result
during measurement.

Chapter 2

Gas Temperatures

The combustion gases of industrial flames are often laden with solid particles at temperatures which may be different to that of the surrounding gas. The present discussion will be confined to the measurement of the gas temperature alone. The equilibrium temperature of the hot junction of a thermocouple inserted into a gaseous stream is a resultant temperature of:

(i) heat transfer by convection between the thermocouple and the gas across the boundary layer;

(ii) heat transfer by radiation between the thermocouple, the gases, suspended particles and the walls of the furnace;

(iii) recovery factor due to transfer of kinetic energy of the gas to thermal energy within the boundary layer of the hot junction;

(iv) heat transfer by conduction along the thermocouple wires when they pass through a temperature gradient which is not negligible.

In practice, the true temperature of the gas can be determined from the thermocouple measurement by one of the following three methods:

(a) Correcting the measured value by using the equations for the errors mentioned above.

(b) Using a system which reduced the errors to a negligible amount.

(c) Using a suction pyrometer in which the temperature of the gas is measured under known conditions and a correction factor, determined from a previous calibration, is applied to the measured temperature.

At the experimental station in IJmuiden, the first method is used for measurement of the temperature of air supplied to the furnace, while the suction pyrometer is used for the measurement of temperatures in the flame. Before examining the suction pyrometer, in detail, the errors that can arise when using a bare thermocouple are discussed, in order that the precautions that need to be taken in the construction of a suction pyrometer can be appreciated.

2.1 Gas temperature measurement with a bare thermocouple

The equations governing the heat exchange between a thermocouple and its environment are discussed in detail in the literature (7)(8)(9), and are summarized briefly below.

Convection

Heat transfer by convection, which we wish to augment, increases with the difference in temperatures T_0 of the gas and T_T of the thermocouple junction (a difference which should be kept as small as possible) and also increases with the convective heat transfer coefficient h_c:

$$Q_c = h_c(T_0 - T_T)$$

The coefficient h_c is generally given in the form of a relation between the Nusselt, Prandtl and Reynolds dimensionless numbers. For combustion gases (Prandtl number approximately 0·7) the relation between the Nusselt number $(Nu) = h_c d/k$ and the Reynolds number $(Re) = \rho V d/\mu$, is (7):

$$(Nu) = (0{\cdot}44 \pm 0{\cdot}06)(Re)^{0{\cdot}5}$$

for a thermocouple placed normal to the flow direction, and

$$(Nu) = (0{\cdot}085 \pm 0{\cdot}009)(Re)^{0{\cdot}674}$$

for a thermocouple placed along the flow direction.

The result is that in the region below the point common to the two equations, *i.e.* for $(Re) < 15\,000$ (Fig. 2.1), the first arrangement is preferable to the second in order to obtain a greater exchange of heat by convection. In addition, h_c increases with velocity but decreases as the diameter of the thermocouple is increased according to the expression

$$h_c = C \cdot V^n d^{-m} \quad \text{with } 0{\cdot}5 < n < 0{\cdot}7 \text{ and } 0{\cdot}3 < m < 0{\cdot}5$$

Also, h_c increases considerably with turbulence so that in some instruments it may be of interest to artificially augment the turbulence.

Velocity of gas

Changes in temperature as a function of velocity are given by the equation of Saint-Venant (for constant enthalpy):

$$\Delta T = \frac{V^2}{2Jc_p}$$

$\Delta T = T_0 - T_V$ is the temperature difference due to loss or recovery of
                  kinetic energy,

$V =$ the gas velocity,

$c_p =$ the specific heat of the gas at constant pressure. Variation with temperature is small and the value remains close to 1 kJ/kg°C for combustion gases.

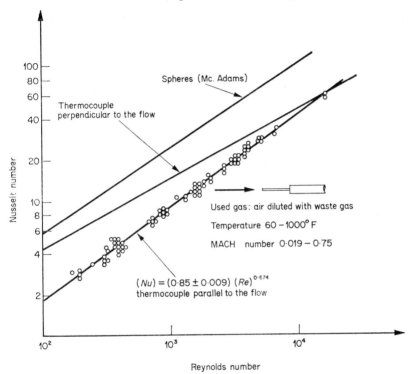

Fig. 2.1 Relation between Reynolds and Nusselt numbers for flow over
the hot junction of a thermocouple

The temperature of the surface of the thermocouple differs from the stagnation temperature as a result of an energy exchange within the boundary layer so that (**10**)

$$\Delta T = \alpha \frac{V^2}{2Jc_p}$$

where α is a recovery factor, with values, for velocities up to Mach 3, according to (**7**), (Fig. 2.2):

$\alpha = 0.68 \pm 0.07$ for thermocouple normal to flow direction
$\alpha = 0.86 \pm 0.09$ for thermocouple along the flow direction

In industrial furnaces gas velocities are generally sufficiently low so that the error, due to the fact that the thermocouple measures the stagnation temperature, is not significant.

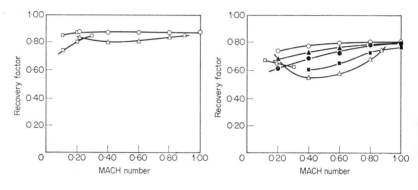

○ from Wehrman
● " Wehrman
▲ " Simmons
□ " Hottel and Kalitiwsky
■ " Glawe, Simmons and Stickney

Fig. 2.2 Recovery factor

Radiation

If σT^4 is the total energy radiated to the thermocouple by the walls and the gas, the heat exchange by radiation Φ_R, which we wish to maintain as low as possible, is given by the expression:

$$\Phi_R = \varepsilon\sigma(T_T{}^4 - T^4)$$

ε being the emissivity of the thermocouple.

Polished metal surfaces have a low emissivity at low temperatures but the emissivity increases rapidly with temperature as well as by oxidation or deposition on the surface (this is also true for the noble metals). Refractory sheaths (Al_2O_3) have, on the contrary, an emissivity that decreases as the temperature increases and they are much less sensitive to chemical action by the gas. In order to reduce Φ_R we require that the temperature T should be as close as possible to T_T. This can be achieved by surrounding the thermocouple with a system of shields with a means to control and modify the temperature so as to bring the temperature of the shields close to that of the thermocouple T_T. MacAdams (9) shows that when $T_T - T$ is small in comparison with T the error can be expressed by:

$$\Delta T = (T_T - T)\frac{h_r}{h_c}$$

where h_r is the radiation transfer coefficient:

$$h_r = \varepsilon(4 + n)\sigma T^3$$

n being a positive number for metal surfaces and negative for refractory surfaces.

It can be seen, therefore, that the error is a function of the temperature difference $(T_T - T)$ and h_r but the error also increases with temperature since h_r increases far more rapidly than h_c as a function of temperature T.

Conduction

For the particular case of a length L of thermocouple wire immersed in a homogeneous flow and immediately emerging into an environment with temperature T_B, the difference E_c between the measured temperature and that of the gas is given by (7)(9)(12):

$$E_c = \frac{T_T - T_B}{L(4h/d\lambda)^{\frac{1}{2}}}$$

with $\lambda =$ thermal conductivity of thermocouple
and $h =$ transfer coefficient between gas and thermocouple.

To reduce the error E_c it is therefore necessary to immerse the thermocouple wires in the flow with as great a length as possible of small diameter and with a low thermal conductivity λ.

Response time of thermocouple

The differential equation obtained by assuming that the heat received by convection heats the thermocouple (neglecting radiation) is given by

$$T_G - T_T = A\frac{dT_T}{dt}$$

where $A = \dfrac{\rho c d}{4h_c}$ with $T_G =$ gas temperature,
and $c =$ specific heat of thermocouple material.

The equation shows that the response time can be increased by increasing h_c and reducing the diameter and specific heat of the thermocouple.

2.2 Suction pyrometer

Principle

In order to obtain directly the true temperature of the gas we wish to find a means of eliminating the various sources of error enumerated in the previous section. This can be achieved by fixing the thermocouple on the axis of a system of shields through which the gas is sucked at high speed, Fig. 2.3 (a). The theoretical and experimental study of Land and Barber **(8)(13)** allows predictions to be made of the performance of the pyrometer as a function of the variables. For turbulent conditions ((*Re*) > 2000) convective heat transfer in a tube, or between two concentric tubes is given by

$$\frac{h_c d}{k} = 0\!\cdot\!20 \left(\frac{\rho V d}{\mu}\right)^{0\cdot8}$$

where d is the internal diameter of the tube or an equivalent diameter for the distance separating two concentric tubes. The coefficient h_c varies only slightly with d so that in order to reduce the dimensions of the instrument while increasing the number of shields, d should be chosen as close as possible to the minimum diameter that allows turbulent flow to be conserved. Moreover, the turbulence is conserved over a certain distance from the entrance to the tube so that, in practice, if the thermocouple is sufficiently close to the entrance the efficiency is not reduced for Reynolds numbers considerably lower than 2000. Since h_c increases with turbulence it is of interest to find means to augment the turbulence by imposing an irregular flow at the tube entry (preferably rough) and placing the thermocouple sufficiently close to the gas entry. Convection is further increased by sucking the gas at as high a suction rate as possible, taking into account that the velocity should be limited by the error which it introduces (Section 2.1).

Error due to velocity. The gas velocity V_G in an industrial furnace varies, in general between 0 and 80 m/s, and the temperature to be measured is that corresponding to the actual velocity and not to the stagnation temperature T_0. For $V_G = 80$ m/s the difference between these two temperatures is:

$$T_0 - T_{V_G} = 80 = \frac{V_G{}^2}{2Jc_\mathrm{p}} = \frac{6400}{2000} = 3\!\cdot\!2°\mathrm{C}$$

If the gases are sucked into the pyrometer at a velocity $V_A = 200$ m/s, the temperature T_m measured by the thermocouple will differ from T_0 by the amount

$$T_0 - T_m = (1 - \alpha)\frac{V_A{}^2}{2Jc_\mathrm{p}}$$

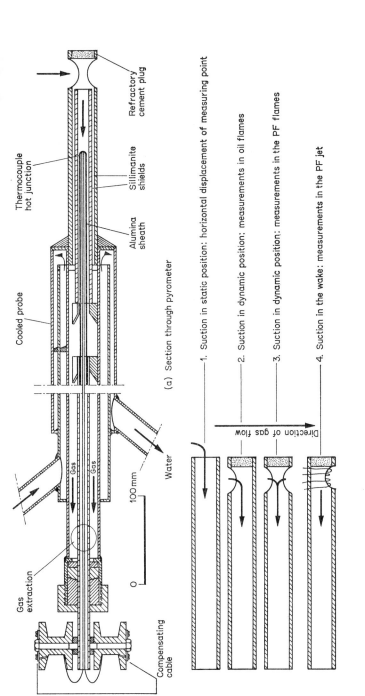

Thermocouple
hot junction

Refractory
cement plug

Sillimanite
shields

Alumina
sheath

Cooled probe

(a) Section through pyrometer

Gas
extraction

Water

Gas

Gas

100 mm

0

Compensating
cable

1. Suction in static position: horizontal displacement of measuring point

2. Suction in dynamic position: measurements in oil flames

3. Suction in dynamic position: measurements in the PF flames

4. Suction in the wake: measurements in the PF jet

Direction of gas flow

(b) External shield

Fig. 2.3 Standard suction pyrometer

Taking $\alpha = 0.85$, as indicated in section 2.1, we obtain

$$T_0 - T_m = 3°C$$

i.e. the thermocouple measures the true temperature at 80 m/s. In general the error due to velocity is given by the equation

$$T_{V_G} - T_m = \frac{(1 - \alpha)V_A{}^2 - V_G{}^2}{2Jc_p}$$

This error is no more than 3°C for $V_A = 200$ m/s; 4.7°C for $V_A = 250$ m/s and 7°C for $V_A = 300$ m/s.

Error due to conduction. It is more simple, from a construction point of view, to place the thermocouple wires along the flow direction, and to achieve a high efficiency it is preferable to use small diameter thermocouple wires such that the conduction error is negligible.

Error due to radiation. The error due to radiation is reduced by increasing the number of screens; increasing their thickness while selecting a material with low emissivity and low conductivity. Land and Barber have taken as reference a steel shield with emissivity close to unity and they give the equations for calculating the equivalence coefficients for other systems as a function of the material properties and geometry (13), Fig. 2.4.

Efficiency. The resultant error depends therefore on the gas temperature, the suction velocity and the construction of the instrument, as well as upon the environmental conditions. These errors are represented by the equilibrium temperature of the thermocouple when no suction is applied, and it is clear that the error will depend upon the differences between this equilibrium temperature and the true temperature of the gas. Land and Barber (15) have defined the efficiency factor E depending upon the pyrometer construction, the suction velocity and the gas temperature but which is independent of the environmental conditions. It is related to the resultant error $T_M - T_G$ by the relation

$$T_G - T_M = (1 - E)[T_G - T_0]$$

where T_G is the true temperature of the gas,
T_0 is the temperature without suction, and
T_M is the measured temperature.

This equation, after rearrangement, gives

$$E = \frac{T_M - T_0}{T_G - T_0}$$

When the value of the efficiency factor E is known for the measuring conditions, the true temperature of the gas can be determined from

$$T_G = \frac{T_M - T_0(1 - E)}{E}$$

In practice the measurement of T_0 is not convenient and it is preferable to find an instrument with an efficiency factor E sufficiently large so that the

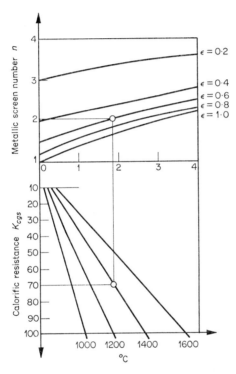

Fig. 2.4 Equivalence of the system of shields as a function of their physical properties

error would be small even under extreme conditions. In the furnace at IJmuiden the value of $T_G - T_0$ does not exceed 400°C for the high temperature conditions. If we accept an error of 10°C the efficiency of the pyrometer would be at least

$$E = \frac{390}{400} = 0.975$$

when functioning under the most unfavourable conditions, *i.e.* at high temperatures. According to Land and Barber the value of E can be determined experimentally with the aid of the following factors (Fig. 2.5):

(i) Form factor for the curve of temperature as a function of suction velocity. This factor "f" is defined for the value of the suction velocity, V_m, by

$$f = \frac{T_{V_m} - T_0}{T_{V_m} - T_{\frac{V_m}{4}}}$$

where T_0 is the temperature measured without suction,

T_{V_m} is the temperature measured with suction at the velocity V_m,

and

$T_{\frac{V_m}{4}}$ is the temperature measured with suction at the velocity $\frac{V_m}{4}$.

The graphs obtained by Land and Barber give the value of the efficiency corresponding to this form factor, taking into account the number of equivalent screens for the pyrometer.

(ii) Form factor for the curve of the response of the pyrometer. A second factor is defined

$$f' = \frac{\tau_0}{\tau_m}$$

where τ_0 and τ_m are the times necessary to achieve equilibrium for the thermocouple with and without suction respectively. As previously the corresponding efficiency is obtained from a graph.

The determination of the efficiency allows the pyrometer to be tested under normal conditions of suction, *i.e.* to measure experimentally the magnitude of the error. One can, therefore, either correct the measured temperature by measuring the temperature without suction, or alternatively, if there is insufficient time to take this measurement the maximum possible error can be determined from an estimation of the temperature difference $T_M - T_0$.

Probe

A suction pyrometer consists, therefore, of a thermocouple protected in general from chemical action by the gas by an impermeable sheath and placed in a system of screens that isolate the thermocouple from the surrounding radiation (Fig. 2.3).

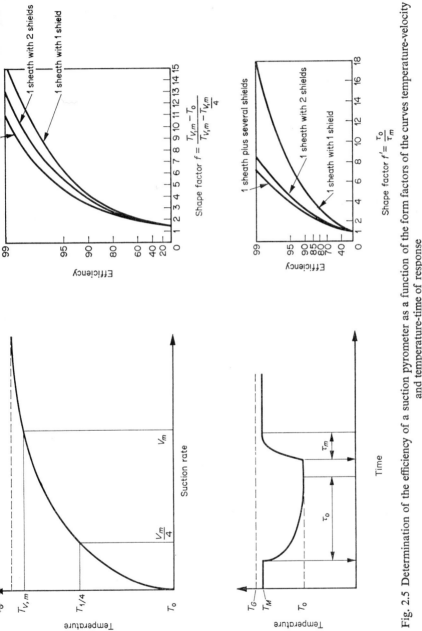

Fig. 2.5 Determination of the efficiency of a suction pyrometer as a function of the form factors of the curves temperature-velocity and temperature-time of response

The gas is sucked at high velocity through the screens and over the sheath by a compressed air or steam ejector, or alternatively by a suction pump.

Thermocouple. The type of thermocouple is chosen according to the temperature range over which the measurements are to be made. The diameter of the wires should be chosen as small as possible but the wires should not be too fragile such that they will break. With platinum it is recommended that the diameter should not be less than 0·5 mm for long lengths of thermocouple and 0·25 mm for shorter lengths of the order of one metre.

In practice it is convenient to choose a thermocouple with a linear characteristic so that the temperature can be read directly from recorder paper graduated in millivolts.

Potentiometers can be obtained with an automatic correction for the cold junction. But, also, thermocouples can be obtained which do not require, in practice, a correction for the ambient temperature (PtRh6%–PtRh30%) (Table 2.1). One can also economize for the long length of

Table 2.1 Error in measurement when the correction for the cold junction is neglected for a thermocouple PtRh6%–PtRh30%

Temperature of the cold junction	Error as a function of the temperature of the hot junction in °C							
(°C)	600	700	800	1000	1200	1400	1600	1800
20	−0·8	−0·7	−0·7	−0·5	−0·5	−0·4	−0·4	−0·4
40	−2·4	−2·0	−1·9	−1·5	−1·3	−1·2	−1·2	−1·2
60	−4·4	−3·7	−3·4	−2·8	−2·5	−2·3	−2·3	−2·3
80	−6·7	−5·6	−5·3	−4·3	−3·8	−3·6	−3·5	−3·5
100	−9·4	−8·0	−7·4	−6·1	−5·3	−5·0	−5·0	−5·0

compensating wires by placing the cold junction in the cooling water of the probe head. The electrical and chemical insulation of the thermocouple wires is made by a sheath and refractory beads (Table 2.2), chosen so as not to have any effect on the thermocouple. At high temperatures, however, perfect insulation does not exist so that a periodic verification of the thermocouple characteristic is necessary to make any corrections (14)(15).

Shields. In choosing the shields it is possible to vary both the composition and the geometry of the shields. Land and Barber (13) have shown that a

Table 2.2 Comparative properties of some refractory oxides as insulants

Insulation, refractory		Purity	Fusion point	Utilization temperature	Coefficient of mean expansion per °C × 10^6	Electrical Resistivity		
						1000°C	1200°C	1400°C
Magnesia	MgO	99·1	2790	2200	12·9	$3·4 \times 10^5$	$2·6 \times 10^4$	$6·4 \times 10^2$
Alumina	Al_2O_3	99·5	2010	1850	8	$\approx 5 \times 10^6$	$\approx 5 \times 10^6$	$\approx 5 \times 10^4$
Zirconia	ZrO_2	99·4	2480	1000	5			
Beryllium Oxide	BeO	99·8	2510	2200	8·1	$4·15 \times 10^5$	$4·7 \times 10^4$	1×10^4
Thorium Oxide	ThO_2	99·5	3200	2800				

shield is equivalent to C metal shields with emissivity 1, according to the equation

$$C = \sqrt{\frac{1}{\varepsilon^2} + f} \quad \text{where } f = 4\sigma T^3 \frac{W}{\lambda}$$

with ε the emissivity of the shield, W the thickness and λ the thermal conductivity (Fig. 2.4).

The shields will therefore be better according as their emissivity and conductivity are lower. This is the case with the white refractories where the emissivity, which diminishes when the temperature increases, can be reduced to a value of 0·2 (**16**), and the thermal conductivity is very low (1 Kcal/h m°C). It may be noted that the dimensions of the refractory particles and porosity have an opposite effect on ε and λ; ε diminishes for small and closely-packed refractory particles, whereas λ increases.

The ideal shield will therefore be made of a very porous refractory covered with a thin layer of dense white refractory. Under certain conditions an acceptable efficiency can be obtained with a large number of metallic shields when the temperatures of the gas and the surroundings are less than 800°C. At high temperatures it is preferable to use refractory materials.

The efficiency of a system of shields can be increased by increasing the surface over which heat transfer by convection takes place.

Different geometries have been compared and it appears that a system of shields moulded into a single block is less efficient than a number of separate concentric shields fitted with longitudinal centralizing sections or, alternatively, an assembly of small tubes (Fig. 2.6). The equivalence between simple cylindrical shields and a shield of complex form is given by the form factor

$$\sigma = \sqrt{\frac{S}{S_0}}$$

where S is the contact surface with the gas, and S_0 is the surface of a cylinder of the same mean diameter as the complex shield.

Surfaces of contact between the shields should be avoided and the centralizing sections should be as pointed as possible.

Since the shields are subjected to sudden changes in temperature, porous refractories will be more satisfactory. It is also possible to use shields with very dense alumina provided the probe is used with precaution.

Suction. The efficiency of the instrument increases with the suction velocity more or less up to velocities of the order of 250 metres per second (Fig.

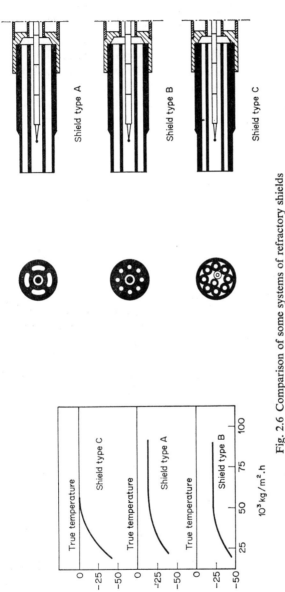

Fig. 2.6 Comparison of some systems of refractory shields

2.7). It is important therefore to control the suction flow rate and to maintain it at a value according to the desired efficiency. The flow rate can be controlled by a venturi or an orifice plate but care must be taken due to the risk of blockage by condensation of vapour, or deposit of particles in the pressure tappings for gases heavily laden with particles. Too high a suction rate can lead to a rapid blocking of the shields, which may, as a result, have to be changed. In order to prevent this blockage it is recommended in practice that the suction velocity should not exceed 100 metres per second. When the risk of blockage by particles does not exist the suction velocity should be maintained above 200 metres per second. The distribution of velocity within each shield depends upon the geometry. One should, in all cases, attempt to increase the flow rate over the thermocouple sheath and arrange that the gas velocity over the sheath is of the order of 250 metres per second.

The flow areas between the shields as well as the suction velocity are conditioned by the suction flow rate that will provide the minimum disturbance to the flow of gases at the measuring point. It is also important that the probe should not alter the local conditions of temperature. It is for this reason that in a gas stream the best position for the screen is that which allows direct and isokinetic suction along the shield, Fig. 2.3 (b). Under these conditions the average temperature of the stream of fluid coming directly from the pyrometer opening is measured. It is preferable to use instruments of small dimension so as to provide minimum disturbance to the local flow, and attempts should be made to remove small quantities of gas at high suction velocities.

Examples of suction pyrometers

At the experimental station at IJmuiden, suction pyrometers of standard form have been used for a number of years for the measurement of temperatures in oil and pulverized coal flames.

More recently, newer pyrometers have been constructed in order to make measurements in coal-laden jets before the flame front where the temperatures are low, but particle concentrations are high. Pyrometers have also been developed for measurements near the hearth of the furnace when the flame is enriched with oxygen and is inclined so as to impinge upon the hearth, *i.e.* conditions of very high temperature and multiple probes offering minimum obstruction.

Measurements in the flame—standard pyrometer (Fig. 2.3) (5)(17). The head of the standard pyrometer consists of a thermocouple, Pt-PtRh10%, protected against chemical action of the gas by a sheath of fused pure alumina. Two concentric shields of sillimanite (60% to 80% Al_2O_3 + SiO_2, with external diameter of the assembly, 28 mm). The head is fitted to the end of a water-cooled probe through which the gases are sucked and the thermocouple wires passed. The gases are sucked by a compressed-air ejector and the flow rate is controlled by a venturi. The thermocouple has a diameter of 0·5 mm and a linear characteristic between 700°C and 1700°C. The cold junction can be placed at the exit of the probe (Land–IJmuiden version) or in the cooling water in probe head (Irsid–Meci version).

The alumina sheath is gas-tight and allows no penetration of combustion gases, including hydrogen, for the whole range of temperatures measured. When the probe was first used with suction and in a "static" position the measured temperature profiles were found to be non-symmetrical in a flame which, in principle, should have been symmetrical (**21**). A lateral hole was then made at 15 mm from the end of the shield in order to allow suction in the "dynamic" position, Fig. 2.3 (b). The cross-sectional area of this hole (18 × 20 mm) was chosen so that the velocity of the gas would be between 20 and 40 metres per second, *i.e.* close to the average velocities encountered in the IJmuiden furnaces. In order to avoid too rapid blockage in flames with pulverized coal, a second lateral hole was made in the shield to allow the large particles to flow directly through the shield without deposition. This extra hole makes the shield more fragile and should only be used when necessary.

On the basis of a study, the hot junction of the thermocouple was fixed at a distance of 40 mm from the entrance of the shield. No significant difference was found in the measured temperature when the position of the thermocouple was varied between 20 mm and 70 mm from the entrance (**18**).

Efficiency and response time. The efficiency of this pyrometer determined from the curve of temperature as a function of suction velocity, Fig. 2.7 (a), is 0·98 for a suction velocity of 250 metres per second and temperature of 1600°C.

The difference between the temperatures with and without suction is seldom more than 400°C and the error, other than that due to the thermocouple itself, is less than 8°C.

The time required to achieve equilibrium is of the order of three minutes for the first measurement and for subsequent measurements, where temperature changes are of the order of only a few 100 degrees, then the response time is of the order of $1\frac{1}{2}$ minutes.

Miniature pyrometer for measurement in the dust-laden jet. In order to analyse in detail the ignition region of the pulverized-coal jet, it was necessary to reduce the size of the standard pyrometer in order to reduce

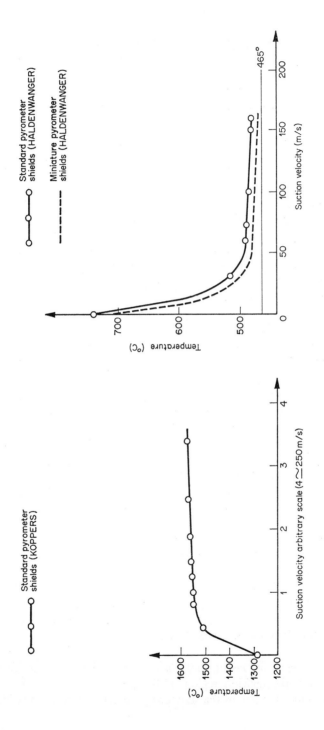

(a) Temperature with suction inferior to gas velocity

(b) Temperature with suction superior to gas velocity

Fig. 2.7 Influence of suction velocity on measured temperature

the disturbance created by the instrument at the measuring point. The most serious disturbance was that of the breaking up of the dust-laden jet and a displacement of the flame front. The flame becomes stabilized in the turbulent zone created by shields which are too large. The required temperature range lies between 80°C and the ignition temperature for various types of solid fuel with high concentrations of particles and radiation from the refractory walls at 1200°C.

Description (Fig. 2.8) (**19**). The miniature pyrometer is reduced to 15 mm in diameter. The thermocouple which is no more than 0·25 mm in diameter is insulated by a pure alumina sheath of external diameter 1·8 mm. The two shields are of sillimanite but less porous and harder than those of the standard pyrometer. They, therefore, can withstand thermal shocks to a lesser degree.

Stainless-steel shields were tried but were found unsuitable for the measuring conditions (Fig. 2.9 (a) and (b)) due to their fragility at high temperatures—softening of the metal and attack by the solid fuel particles.

In order to avoid blockage the suction opening was placed in the wake of the instrument where the particle concentrations are very low. Experience has shown that this arrangement does not introduce any significant error to the measurement, and that the location of the measuring points could be determined satisfactorily under the conditions where the suction velocity through the opening was close to that of the gas flow (**19**). By these means the blockage is completely eliminated, except in those regions where the velocity is low and the turbulence intensity is high, *i.e.* when the inertia of the particles is insufficient to escape suction.

Efficiency and response time. The efficiency of this pyrometer at 500°C for a suction velocity of 160 metres per second is equal to 0·995. The measured temperature is thus within 2°C of the true temperature, except under the conditions where solid fuel is deposited and burnt on the upstream face of the shield, or alternatively when the flame front is brought back to the shield. Under these conditions the measurements are so clearly in error that they can be easily identified.

The distance between the shields was between 1 mm and 1·5 mm so that the flow was laminar at 500°C for the suction velocities used. The reduction in efficiency of the pyrometer was not tested experimentally at this temperature but only for the much higher temperatures at 1500°C. Moreover the efficiency remains satisfactory up to 1700°C where $E = 0.951$.

The response time of this pyrometer is considerably shorter than that of the standard pyrometer. Equilibrium is achieved in less than one minute for large variations in temperature, and in a few seconds for a variation less than 100°C, so much so that it is even possible to partially follow the rapid fluctuations in temperature. The blockage of the shields was

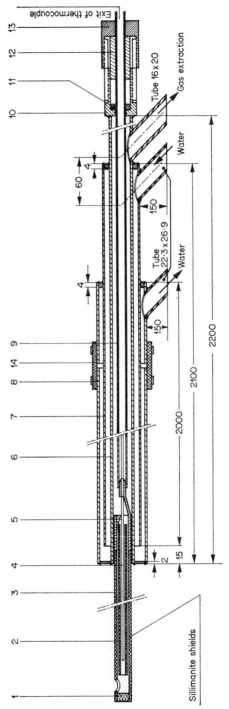

Fig. 2.8 Miniature pyrometer

Exit of thermocouple

Tube 16 x 20

Gas extraction

Water

Tube 22·3 x 26·9

Water

Sillimanite shields

eliminated and the response time of the pyrometer reduced so that measurements of temperature could be made in the dust-laden section of the jet four times more rapidly than measurements in the flame with the standard pyrometer.

Multiple pyrometer for measurements in the vicinity of the hearth. This instrument was developed for the measurement of the temperature profile

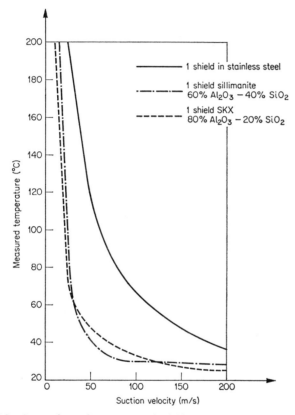

Fig. 2.9a Comparison of some types of shield for measurement of temperatures in a cold jet under conditions of strong radiation

in the vicinity of the hearth heated by a flame, enriched with oxygen, impinging on the hearth (**20**). It consists of four pyrometers mounted in a vertical position at four different heights on a probe, which is introduced along channelled grooves in the hearth (Fig. 2.10). Each pyrometer, which is of the miniature type, consists of a sheath and two shields. The probe had to be miniaturized in order to provide the minimum disturbance to

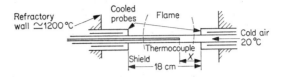

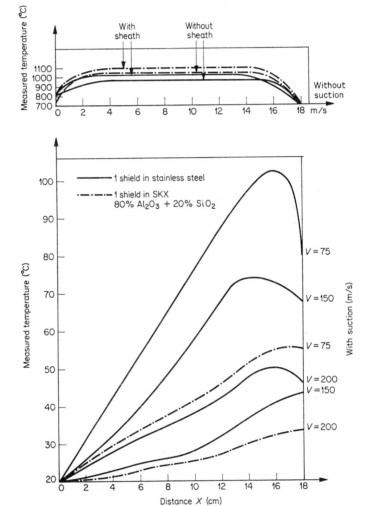

Fig. 2.9b

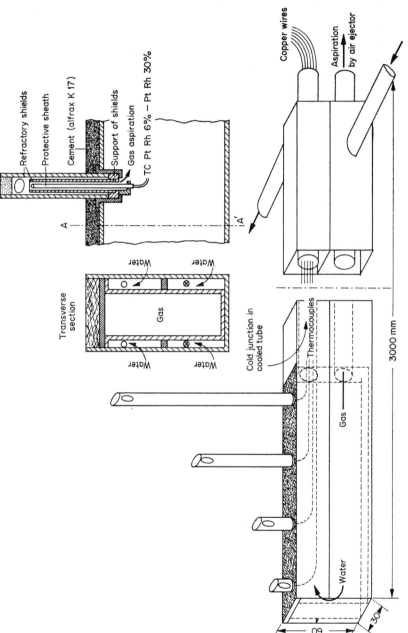

Fig. 2.10 Quadruple suction pyrometer

the flow in the immediate vicinity of the hearth and also to have a short
response time and have the possibility of being mounted on a probe of
small dimensions. This arrangement allowed a reduction in the time required to position and manœuvre the probes under conditions where
it is important to take a large number of measurements.

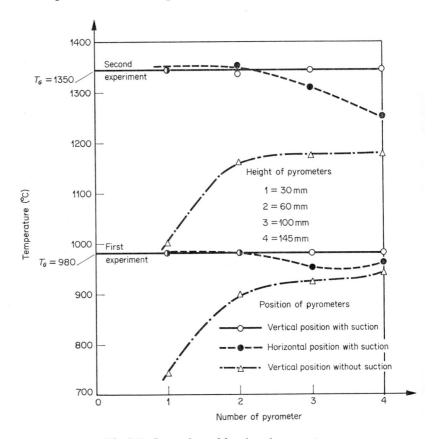

Fig. 2.11 Comparison of four hearth pyrometers

Description (**21**). Because of the oxygen enrichment in the flame maximum
temperatures were of the order of 1850°C (**20**). A thermocouple of
PtRh6%–PtRh30% was therefore chosen, which allowed the placing of
the cold junction in the cooling water of the probe and to include the
correction for the cold junction (Table 2.2). The diameter of the thermocouple was 0·25 mm. In order to provide resistance against the high temperatures and to allow the reduction in the dimensions of the probe

without increasing its fragility shields of pure alumina were used. Thermal conductivity of these shields is considerably higher but the emissivity is less than the porous refractory materials used for the shields in the pyrometers described previously. This choice is justifiable at high temperatures where the emissivity, reduced to 0·2, becomes the predominant factor in the search for a high efficiency. The independent shields do not have centralizing grooves. Experience has shown that the omission of the grooves had no effect when the pyrometer was used in a vertical position, but this would not be the case for the longer pyrometers if these had been used in a horizontal position (Fig. 2.11).

In order to simplify the assembly and to achieve a simultaneous equilibrium of the four pyrometers the suction was arranged so that the gas from each pyrometer entered a single, large-volume vessel, in which the pressure was reduced by the ejector. Despite the effect of equalization due to the large pressure drop caused by the support at the base of the shields the suction velocity was larger in the shorter shields and compensated, therefore, for the cooling due to the proximity of the probe. This assembly is only suitable in flames where there is no risk of frequent blockage, which would not be revealed by the measurement of the flow rate but only by the drift in the measurement.

Efficiency and time response. According to the results of Land and Barber (**13**), taking into account the emissivity of the refractory as approximately 0·2 and also the effect of convection on the outer walls of the shield due to the velocity of gases in the furnace, the efficiency of this system of shields was found to be more than 0·90. The calculation of the error is difficult, however, due to the lack of precision in the determination of the temperature without suction—the thermocouple in the shortest pyrometers are slightly cooled by conduction when there is no suction, and this cooling is considerably reduced when the support for the shield and the thermocouple wires are exposed to the hot gas stream.

A comparison carried out in a furnace at 1000°C and 1350°C (Fig. 2.11) showed that the four pyrometers indicate the true temperature within a few degrees for an average suction velocity of 115 metres per second (**21**). For the maximum temperatures measured (1850°C) it is probable that reductions were lower by some tens of degrees. For variations in temperature of several hundreds of degrees the response time is about 30 seconds. Due to the elimination of the measurement of the cold junction and the achieving of simultaneous equilibrium of the four pyrometers, eight times as many measurements were made with this pyrometer as with a standard pyrometer in the same time. This results in a considerable reduction in the time and cost of the measurement without any reduction in precision.

2.3 Instruments requiring calibration

Even though these instruments have not been used to any large extent at IJmuiden it is useful to mention them since they have some particularly interesting characteristics for certain applications.

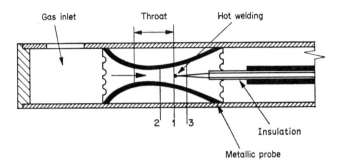

(a) Scheme of apparatus

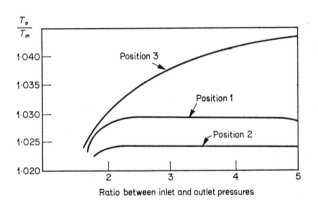

Ratio between inlet and outlet pressures

(b) Influence of the position of hot welding in the throat

Fig. 2.12 Sonic pyrometer

Sonic pyrometer (**11**)

The hot junction of the thermocouple is placed in the throat at the head of the suction tube where the gases are flowing at the speed of sound (Fig. 2.12). The measured temperature T_m is related to the total temperature of the gas T_0 by the relation

$$T_0 = kT_m$$

where

$$k = \frac{\gamma + 1}{2 + r(\gamma - 1)}$$

$r =$ recovery factor, and
$\gamma =$ ratio of specific heats of the gas.

The value of k, measured at ambient temperature, is corrected for high temperatures by using the law of the variation of γ as a function of this variable.

Due to the high suction velocities the error due to radiation, despite the presence of only one shield, is negligible and the response of the pyrometer is very rapid.

This pyrometer is in general constructed of "inconel" and can be used at temperatures up to the order of 1200°C. It can also be manufactured in alumina.

Suction pyrometer with cooled shield (22)(23)

The gases are sucked over a thermocouple placed in a water-cooled probe. If the suction flow rate is maintained constant as well as the temperature of the cooling water, the temperature measured by the thermocouple only depends upon the temperature of the gas at the probe entrance. Experience has shown that for a suitable and reproducible arrangement of the thermocouple in the probe, the calibration curve obtained for one pyrometer was also valid for other similar pyrometers. It is necessary, therefore, only to make the calibration once. Such a system has the advantage of being robust and could be used at very high temperatures. On the other hand it has the disadvantage of requiring a calibration which is often difficult to carry out.

2.4 Venturi pneumatic pyrometer

The venturi pneumatic pyrometer is based upon a different principle than that of the suction pyrometer. It was invented more than 70 years ago but despite its many advantages, it was not developed because of the technological difficulties involved in its construction and the lack of information about certain sources of error (11). Due to the inability of the standard suction pyrometer to measure temperatures that are very high or to make measurements in gases heavily laden with dust, some 15 years ago studies were restarted on this type of pyrometer. As a result of the studies made by Godridge, Jackson and Thurlow at BCURA a number of probes have been made and are now sold commercially by Land who are continuing the development of the instrument (24)(25).

Principles

The law of conservation of mass flow Q of a gas sucked in two venturis, 1 and 2, in series is given by (Fig. 2.13),

$$Q = \rho_1 s_1 v_1 = \rho_2 s_2 v_2$$

which together with

$$\Delta p_1 = K_1 \frac{Q^2}{\rho_1} \quad \text{and} \quad \Delta p_2 = K_2 \frac{Q^2}{\rho_2}$$

gives

$$\frac{\Delta p_1}{\Delta p_2} = \frac{K_1}{K_2} \frac{\rho_2}{\rho_1}$$

and taking into account the perfect gas law, $p/\rho = R.T$, which can be written as

$$\frac{\rho_2}{\rho_1} = \frac{p_2}{p_1} \frac{T_1}{T_2}$$

we obtain finally

$$\frac{\Delta p_1}{\Delta p_2} = \frac{K_1}{K_2} \times \frac{p_2}{p_1} \times \frac{T_1}{T_2} \qquad (2.1)$$

with K = venturi constant
  Δp = differential pressure of the venturi
  p = gas pressure at the venturi inlet
  T = gas temperature at the venturi inlet

If the first venturi, placed in the probe head, receives directly the furnace gas which, after cooling in the probe, passes through the second venturi, the initial gas temperature is given by the relation

$$T_1 = k \frac{\Delta p_1}{\Delta p_2} T_2 \qquad (2.2)$$

where k is the constant of the instrument based on the assumption that the ratio $\Delta p_1/\Delta p_2$ varies by a negligible amount.

This is the principle of the Land–BCURA probe, tested at IJmuiden (25) in which the differential pressures were measured by a pressure transducer and the gas temperature T_2 at the exit from the second venturi was measured by a resistance thermometer. An analogue computer is used so as to give a direct reading of the temperature on a potentiometric recorder fitted with two ranges of measurement, 0°C–250°C and 0°C–2500°C (Fig. 2.13). The constant k is determined by sucking the gas at ambient temperature ($T_1 = T_2$). The advantages of this probe are the following:

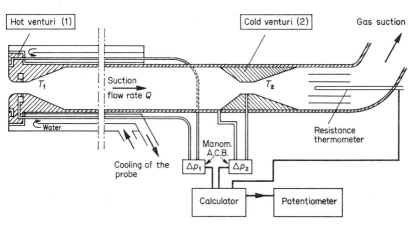

Fig. 2.13 Schematic diagram of two venturi pyrometers of Land

(i) Wide range of temperature measurement, *i.e.* 0°C–2500°C.
(ii) Response time of the order of a few seconds allowing measurements of rapid temperature fluctuations in the flame.
(iii) Insensitivity to blockage in gases heavily laden with dust.
(iv) Robustness and ease of maintenance allowing continuous operation.
On the other hand it has the disadvantage of being sensitive to a number of important sources of error which it is necessary to take into account, in order to predict the performance of the instrument as a function of measuring conditions.

Sources of error

The sources of error can be divided into two categories depending upon whether they result from the principles or the practical application of the instrument.

Errors arising from the method of measurement. (1) Equations (2.1) and (2.2) refer to gases obeying the perfect gas law (low pressures) and flowing under turbulent conditions ((Re) > 2200). These conditions are generally obtained in the instrument. On the other hand in equation (2.2) it was assumed that the ratio of the pressures $\Delta p_1/\Delta p_2$ remains constant and that the gas is incompressible. This is not strictly true, and account must be taken of this by considering the changes in the coefficient k at the time of measurement in relation to the coefficient k' of the calibration.

$$\frac{k}{k'} = \left(\frac{p_{1c}}{p_{1c}'} \cdot \frac{p_{1a}'}{p_{1a}} \times \frac{p_{2a}}{p_{2a}'} \times \frac{p_{2c}'}{p_{2c}}\right)^{2/\gamma} \cdot \frac{p_{1a}}{p_{1a}'} \times \frac{p_{2a}'}{p_{2a}}$$

The indices a and c represent, respectively, the values upstream and at the throat of the venturi, and γ is the ratio of specific heats. The resulting error could be relatively important as shown by the example given by Land (**26**):

Values at calibration	Values during measurement
$\Delta p_1' = \Delta p_2' = 254$ mm WG	$\Delta p_1 = 254$ mm WG; $\Delta p_2 = 85$ mm WG
$p_1' = 10\cdot160$ mm WG	$p_1 = 10\cdot160$ mm WG
$p_2' = 10\cdot010$ mm WG	$p_2 = 9\cdot950$ mm WG
$T_1' = T_2' = 300°K(\gamma = 1\cdot4)$	$T_2 = 600°K; T_1 = 1800°K$

therefore

$$\frac{k}{k'} = 0\cdot978 \times 1\cdot005 = 0\cdot983 \text{ and } T_1 = 1800 \times 0\cdot983 = 1770K$$

The absolute error is therefore 30°C in excess and the relative error

$$\frac{\Delta T_1}{T_1} = \frac{30}{1770} = +1\cdot7\%$$

(2) The equations are only valid for a mixture of gases in which the molecular weight and the mass do not change between the two venturis. This can occur due to:

(*a*) dissociation and recombination. Changes in the value of the specific mass of the mixture occur when there is dissociation, and this results in an error in the ratio of the differential pressures of the two venturis and thereby on the temperature measurements. The effect of dissociation does not become significant until very high temperatures are reached, as is shown in the following table (**31**) for a gas mixture of $25 \cdot 73\% CO_2$, $5 \cdot 23\% H_2O$ and $69 \cdot 04\% N_2$.

Temp. (°C)	Dissociation (%)		$\dfrac{(\sum_i \gamma_i m_i') \text{ at } T°C}{(\sum_i \gamma_i m_i) \text{ at } 30°C}$
	CO_2	H_2O	
1700	2·2	0·5	99·78
1900	6·9	2·9	99·24
2100	15·0	8·0	98·39
2300	26·9	15·7	97·10
2500	42·9	26·5	95·39

γ_i and m_i are the specific weight and the molecular mass respectively of the component in the mixture of gas j.

(*b*) as a result of chemical reaction when the gases are not fully quenched at the exit from the first venturi so as to prevent any further reactions. In particular when combustible particles are present in the gas the error cannot be negligible and it may be necessary to control the composition of the gas at the entry and exits from the instruments (**27**).

(*c*) as a result of condensation it is important not to cool the gaseous mixture below the dewpoint of the gas component which is the first to condense. This may cause complications in the design of the probes.

(3) Finally, changes in the flow conditions, due to variation of the Reynolds number and heat transfer with gas temperature, can result in the constant k of the instrument being a function of this temperature and thereby a systematic error is introduced in the measurement.

At high temperature the Reynolds number of the flow in the tube becomes less than 10^4. Under these conditions the coefficient of pressure drop of the instrument, $\Delta p_1 / \Delta p_2$ and the venturi constants (Fig. 2.14) vary. The heat transfer to the wall of the tube also changes the velocity profile in the boundary layer and thereby the pressure loss coefficient.

Finally, as the presence of particles affects the differential pressure in the venturis, the specific mass needs to be determined by taking into account that of the gas and that of the mixture. Experience has shown that the particle concentration has to be very high before this error becomes significant (**28**).

Errors due to the probe and assembly. (i) The constants and the dimensions of the venturis depend upon the temperature of the gas. The venturi constants vary as a function of the Reynolds number particularly at low

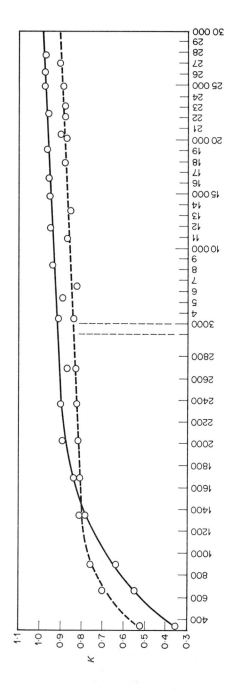

Fig. 2.14 Variation of venturi constants as a function of Reynolds number

– – – – Hot venturi   ——— Cold venturi

values (Fig. 2.14) and the dimensions can change under the effect of thermal expansion and by deposition of particles. The venturi at the head is sensitive to change in Reynolds number and according to the results of a study made in Germany the use of an orifice plate gives better results.

(ii) The cooling which it is necessary to provide for the probe causes the gases to be partially cooled as they enter the throat of the first venturi. In the probes which have been built this cooling is considered to be responsible for the systematic negative error obtained with this type of pyrometer (**29**). On the other hand this influence has been considered to be small (**26**).

(iii) The pneumatic pyrometer is designed for measurements under stationary conditions. When the gas is in motion the magnitude and direction of the velocity influences the measurement of the venturi in the probe head when suction does not take place along the fluid flow streamlines. This results in the measurement of false temperature fluctuations as a result of the turbulence of the flow from the probe, and also a significant error (-3% for $V = 6$ metres per second and $\alpha = 45°$) (**30**) depending upon the magnitude and direction of the flow velocity.

(iv) Finally, there are the errors introduced by the measuring instrument itself which are due to inaccuracies and deviations from linearity of the pressure transducers, the resistance thermometer, and the analogue computer. The magnitude of these errors can be determined by calibration and the most favourable measuring ranges can be selected. Land have stated that for their apparatus the errors do not exceed $1\cdot5\%$. It should be added that the suction flow rate must be maintained constant during the measurement because of its influence on the Reynolds number and due to the non-simultaneous measurements of the three measurements on which basis the temperature is calculated A suction system which is stable is therefore necessary and for this reason a fan is used in preference to an ejector.

Accuracy of measurement. Even though the possible sources of errors are many, many of them do not arise except under very particular conditions or, alternatively, they are systematic and may therefore be corrected by calculation or by calibration. There remains, however, an inaccuracy of $1\cdot5\%$ on the absolute temperature arising from the measurements of the three variables in the basic equation. The calibration cannot be carried out easily and when no calibration is available temperatures are measured, in general, too low with a precision, depending upon the measuring conditions, of approximately 5%. It can be seen therefore that this pyrometer is substantially less accurate than a suction pyrometer correctly constructed, under normal conditions of use but it has, on the other hand, the advantage of maintaining this precision under difficult measuring conditions (very high temperature, gases laden with dust) when other instruments cannot be used.

Response time

The measurement of pressure is practically instantaneous so that the response time of the pyrometer depends essentially on the resistance thermometer. Two characteristic examples are given by Land (**26**).

For the first measurement after insertion in the furnace and presuming that the probe has been calibrated at the ambient temperature and that the probe is introduced into the furnace at a point where the gases are at a temperature T_1,

$$T_1 = \frac{\Delta p_1}{\Delta p_2} \times \frac{\Delta p_2'}{\Delta p_1'} \times \frac{T_2}{T_2'} \times T_1' \text{ with } T_1' = T_2' = 300\text{K}$$

If in the furnace $\Delta p_1/\Delta p_2 = 4$ and $T_2 = 450$K the measurement of T_1 should be $T_1 = 1800$ K, but we have as a function of time (**26**):

Time (s)	Indicated temp. T_2(K)	Indicated temp. T_1(K)	Error (°C)
½ to 1	300	1200	−600
6	395	1580	−220
15	435	1740	−60
30	449	1796	−4

In practice condensation will take place before the instrument achieves thermal equilibrium and it is therefore necessary to wait for a longer period of several minutes in order to obtain the correct measurement.

For measurements under conditions where there is only a small change in temperature, the changes from $T_2 = 450$ K, $\Delta p_1/\Delta p_2 = 4$ and $T_1 = 1800$ K to $T_2 = 455$°C, $\Delta p_1/\Delta p_2 = 4\cdot175$ and $T_1 = 1900$ K are as follows:

Time (s)	Indicated temp. T_2(K)	Indicated temp. T_1(K)	Error (°C)
2	450	1879	−21
6	453·2	1892	−8
15	454·5	1879·5	−2·5
60	455	1900	0

The response time is therefore very short in comparison with the standard suction pyrometer and it will be even shorter when the variation in T_2 is less. This is an important advantage of the pneumatic pyrometer. Since the measurement with the Land pyrometer is, however, influenced by the magnitude and direction of the velocity, the turbulence of the flow at the measuring point results in a fluctuation of the reading so that, at times, it is difficult to have a precise indication of the true fluctuation of the temperature.

Probes

There are two types of pneumatic pyrometer depending upon whether the gas is sucked at a low velocity (approximately 50 metres per second), or at the velocity of sound, but in all cases the essential part of the instrument is in the probe head where the principal sources of error occur. The historic development of the instrument and a review of the many different versions of the instrument are given in (11) and we shall discuss as an example the characteristics of the two probes that have been specially designed for measurements of gas temperatures in industrial furnaces.

Land-BCURA probe (Fig. 2.15). The Land-BCURA probe uses venturis for the measurement of flow rate. The hot venturi can be mounted in different positions according to the range of temperature to be investigated. Version (a) consists of a detachable venturi that can be removed for cleaning and is fixed to a central tube. The gases are cooled indirectly in order to prevent condensation ($T_2 > 150°C$). This system has the disadvantage that the central tube, when expanding diametrically at the commencement of suction is blocked at approximately 50 cm from the hot venturi so that longitudinal expansion takes place not only in the backward direction but also in the forward direction. The venturi emerges from the cooled probe so that the cooling is no longer assured and the pressure tappings become exposed (Fig. 2.16). This may be eliminated either by separating the hot venturi from the central tube or by adding a device that forces expansion to take place backwards.

The version (b) is very robust and is recommended when the gases are at very high temperature and the indirect cooling of the venturi is inefficient. When the probes are introduced laterally with respect to the flame axis, suction takes place in the "static" position so that there is a significant effect of gas velocity on the measured temperature (Section 4.2) and on the position of the effective measuring point. A probe has been constructed at IJmuiden for suction in the "dynamic" position (Fig. 2.17).

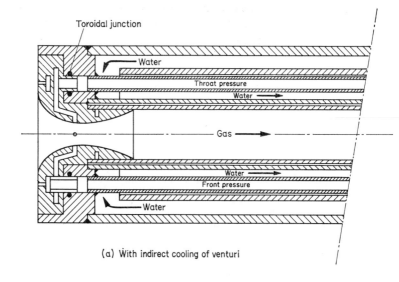

(a) With indirect cooling of venturi

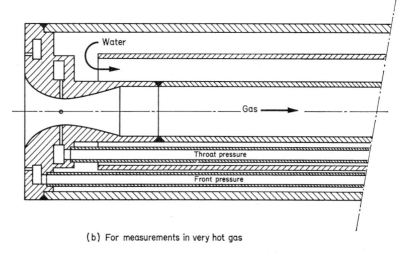

(b) For measurements in very hot gas

Fig. 2.15 Land pneumatic pyrometer—assembly of hot venturi

Kahnwald probe (Fig. 2.18). In this probe the gas flow rate is measured by orifice plates in quarter circles so that the constant does not vary for Reynolds numbers larger than 3000. The cooling of the orifice plate in the probe head is direct and experience has shown (**29**) that this results in a systematic error (reading too low); an error that can be corrected by calibration, taking into account that it is not a function of the gas temperature. Once this correction has been made, it has been established in comparison

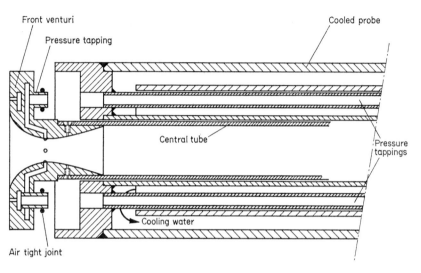

Fig. 2.16 Detachment of hot venturi under the effect of expansion of
the central tube

with a suction pyrometer, that the error does not exceed $\pm$ 1·5% of the absolute temperature in a gas flow of low velocity, lightly laden with dust and chemically stable, *i.e.* in favourable conditions.

Conclusion

The pneumatic pyrometer which is clearly considerably less precise than a standard pyrometer is only of interest for measurements in gases where the temperature is very high or heavily laden with dust.

It is important to select the best probe for the particular measuring conditions.

As regards accuracy it can vary from $\pm 10\%$ in the most unfavourable conditions to $\pm 1\cdot 5\%$ under optimum conditions.

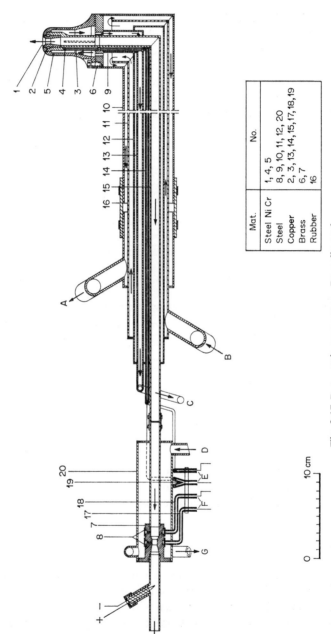

Mat.	No.
Steel Ni Cr	1, 4, 5
Steel	8, 9, 10, 11, 12, 20
Copper	2, 3, 13, 14, 15, 17, 18, 19
Brass	6, 7
Rubber	16

0   10 cm

Fig. 2.17 Pneumatic pyrometer—Pengelly probe

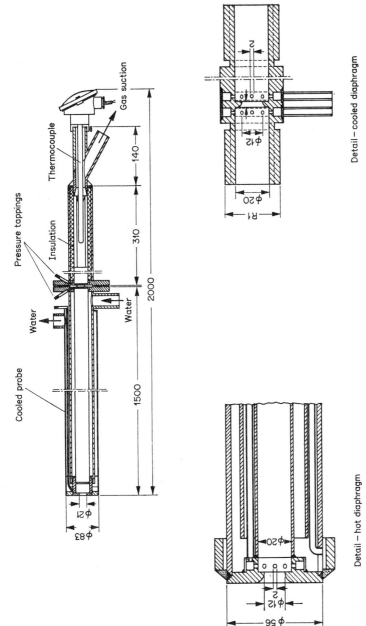

Fig. 2.18 Pneumatic pyrometer—Kahnwald probe

Gas suction

Thermocouple

Pressure tappings

Insulation

Water

Water

Cooled probe

φ21

φ83

140

310

2000

1500

Detail – cooled diaphragm

φ20

φ12

2

R1

Detail – hot diaphragm

φ56

φ20

φ12

2

Bibliography

1. LOISON, R. *Chauffage Industriel et Utilisation des Combustibles. Tome I et II.* Cours à l'Ecole Nationale Supérieure des Mines de Paris. J. et R. Sennac, 1959.
2. SURUGUE, J. and BARRERE. *Combustion.* Librairie Polytechnique Béranger, 1963.
3. *Temperature—Its measurement and control in science and industry.* Volume III, Part 2, Section I, 24. Reinhold Publishing Corporation, 1962.
4. BURTON, J. *Pratique de la mesure et du contrôle dans l'industrie. Tome II.* Dunod, 1959.
5. RIBAUD, G. *Etude de Pyrométrie pratique.* Collection A.N.R.T., 1959.
6. CHEDAILLE, J. and BRAUD, Y. Equipements, méthodes et instruments nouveaux mis en service à la Station d'IJmuiden en 1964 et 1965. *6 ème Journée d'Etudes sur les Flammes,* Paris, November, 1965.
7. *Temperature—Its measurement and control in science and industry.* Volume III, Part 2, Section IV, Article 52. Reinhold Publishing Corporation, 1962.
8. LAND, T. and BARBER, R. Suction pyrometer in theory and practice. *Journ. of Iron and Steel Inst.,* **184,** 289, November 1956.
9. MCADAMS, W. H. *Heat Transmission.* McGraw-Hill, New York, 1954.
10. SCHLICHTING, H. *Boundary Layer Theory.* McGraw-Hill, 1968.
11. *Temperature—Its measurement and control in science and industry.* Volume III, Part 2, Section IV, Articles 53–55–56–57. Reinhold Publishing Corporation, 1962.
12. BERTODO, J. Thermocouple for the measurement of gas temperature up to 2000°C. *Proc. Inst. Mech. Eng.,* Vol. 177, No. 22, 1963.
13. LAND, T. and BARBER, R. The design of suction pyrometer. *Trans. Society Instr. Techn.,* Vol. 6, No. 3, September, 1954.
14. SOCIETE MECI. Etalonnage des couples thermoélectriques. *Revue pratique de contrôle industriel,* No. 5, April, 1963.
15. THOMAS, D. B. A furnace for thermocouple calibrations to 2200°C. *Journal of Research of the National Bureau of Standards,* Vol. 66C, No. 3, July–September, 1962.
16. MICHAUD, M. *Facteurs d'émission d'oxydes métalliques et réfractaires à haute température.* Thèse de Doctorat, 1951.
17. KISSEL, R. R. *Appareils de mesure actuellement utilisés ou en cours de mise au point pour l'étude des flammes.* à IJmuiden. Doc. nr. F 72/a/4, January, 1960.
18. HUBBARD, E. H. and RIVIERE, M. J. M. *Combustion Mechanism trials Nr. II. Suction Pyrometer.* Doc. nr. F 31/a/1, January, 1954.
19. CHEDAILLE, J. and BRAUD, Y. *Essais préliminaires à C-12. Choix des flammes et mise au point des instruments.* Doc. nr. F 32/a/34, January, 1966.
20. CHEDAILLE, J. *Mesure du transfert de chaleur dans le cas de flammes incidentes d'huile avec injection d'oxygène pur. 6ème Journée d'Etudes sur les Flammes,* Paris, Doc. nr. K 20/a/24, November, 1965.
21. BRAUD, Y. *Instruments utilisés durant les essais 0–15.* Doc. nr. F 72/a/10, July, 1966.
22. ATKINSON, P. G. *The measurement of gas stream temperature in industrial appliances.* Gas Council, London, Research Communication GG 33, November, 1956.
23. ATKINSON, P. G. and HARGRAVES, J. R. *The measurement of gas stream temperature in industrial appliances: A suction pyrometer for temperature above 1100°C.* Gas Council, London, November, 1956.
24. LAND, T. Pyrometers Limited, Sheffield, England. Summary of a meeting held at Dronfield on August 27th, 1964, to discuss the design and operation of the Land venturi pyrometer.

25. BRAUD, Y. *Pyromètre pneumatique.* I.F.R.F. Rapport Interne No. 49.
26. LAND, T. Pyrometers Limited, Sheffield, England. Operating instructions for pneumatic pyrometer.
27. THURLOW, G. G. Doc. nr. F 32/a/9—CMC I, August, 1957.
28. HOLLAND, R. E., JACKSON, R., THURLOW, G. G. The behaviour of the venturi pneumatic pyrometer in industrial furnaces. *Journ. Inst. of Fuel*, April, 1960.
29. KAHNWALD, H. Entwicklung eines neuen Absaugethermoelementes zur Messung hoher Gastemperaturen. *Archiv für das Eisenhüttenwesen.* 34, No. 9; pp. 673–678, September 1963.
30. GODRIDGE, A. M., JACKSON, R. and THURLOW, G. G. The venturi pneumatic pyrometer. *Journal of Scientific Instruments*, Vol. 35, March, 1958.
31. SPIERS, H. M. *Technical data on fuel*, 6th ed. British Committee World Power Conference, 1950, p. 440.

Part 2

Measurement of Heat Transfer

Introduction

In industrial furnaces and boilers where turbulent diffusion flames are burning, the heating of the load and the walls takes place simultaneously by convection and by radiation, so that a thorough analysis of the phenomena requires not only the determination of the overall exchanges but also the separation of the respective contributions of the two modes of heat transmission. This primary objective is achieved by means of various heat flow meters:

i) ellipsoidal total radiation heat flow meter,
ii) conductivity flow meter,
iii) circulating water flow meter,
iv) ONERA flow meter, etc.

A knowledge of the heat flow received by convection $\Phi_c = h_c(T_G - T_S)$ enables the coefficient of forced convection h_c to be calculated, this being the only unknown since the temperature of the gas T_G and that of the receiving surface T_S are directly measurable (Section 3.1). This coefficient h_c, is characteristic of the gaseous flow in the region of the surface, i.e. for a given surface condition and geometry, of the flame alone. On the other hand the flow of heat received by radiation, which is generally much more important, originates both from the flame and from the enclosure surrounding it, so that the determination of the radiation of the flame itself must be carried out in another way. This second objective is achieved by means of a total radiation pyrometer, the measurements of which serve in particular for calculating the total hemispherical emissivity of the flame ε_F as well as its temperature T_G.

3

Measurement of the Heat Transferred to a Load

3.1 General introduction

The heat exchange balance between a unit surface element of the load and the combination of the flame and the furnace walls can be written:

$$\Phi_T = \alpha_s \Phi_R - \varepsilon_s \sigma T_s^4 + h_c(T_G - T_S) \tag{3.1}$$

where: Φ_T = heat flow absorbed by the surface and used to heat the load by conduction, per unit surface area

ε_s = total emissivity of the surface, which is a function of the state of the surface and the temperature (1)

α_s = total absorptivity of the surface. If the surface is grey $\alpha_s = \varepsilon_s$

T_S = absolute temperature of the surface

T_G = absolute temperate characteristic of the gas in the region of the surface

h_c = coefficient of convective heat transfer, which is a function of the physical characteristics of the gas and the flow, as well as of the temperature level

Φ_R = flow of heat incident by radiation on unit surface area

Φ_c = flow of heat incident by convection on a unit surface area,

$$\Phi_c = h_c(T_G - T_S)$$

It is assumed in equation (3.1) that the temperatures T_S and T_G are known (see Part 1 for temperature measurements), as well as the absorptivity α which is nearly always equal to the emissivity ε_s, and is found either from Tables or is measured as shown later. Since the measurement of the two variables Φ_T and Φ_R can be carried out by replacing the load element in question by heat flow meters (the ellipsoidal heat flow meter gives Φ

directly, and the others give Φ_T corresponding to a particular surface temperature), Φ_c can be found by calculation.

Before going on to a description of the apparatus we feel it would be useful to define briefly the principal methods of calculation used to evaluate the measurements.

(i) At IJmuiden we use successively the ellipsoidal total radiation heat flow meter and the conductivity flow meter (2). The first gives Φ_R and the second

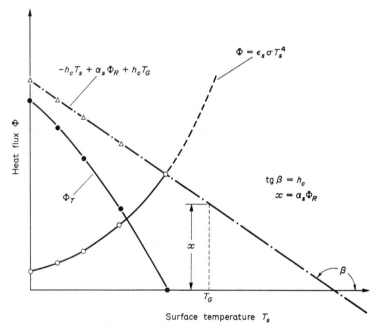

Fig. 3.1 Graphical determination of convection and radiation

Φ_T for a particular surface temperature T_S. The only remaining unknown in equation (3.1) is, therefore, Φ_c:

$$\Phi_c = h_c(T_G - T_S) = \Phi_T + \varepsilon_s \sigma T_s^{4} - \varepsilon_s \Phi_R$$

The coefficient h_c calculated in this way is applicable to the load by using a correction (3) which allows for the difference in the surface temperature between the actual load and the heat flow meter.

(ii) Another widely employed method consists in using several conductivity heat flow meters of differing design in such a way as to obtain the Φ_T corresponding to different surface temperatures. Equation (3.1) written in the form

$$\Phi_T(T_s) + \varepsilon_s \sigma T_s^{4} - \alpha_s \Phi_R = -h_c T_s + h_c T_G$$

shows that the sums of the quantities on the left-hand side should follow a straight line when the temperatures T_S are plotted along the axis of abscissas (Fig. 3.1). The slope of the line is $(- h_c)$ and the ordinate for $T_s = T_G$ is zero (it is assumed that the variation of h_c and α_s as a function of temperature is negligible). This second method is equally applicable with the Onera heat flow meter alone (4), the surface temperature of which can be raised to different levels by electrical heating.

3.2 Hollow ellipsoidal total radiation heat flow meter

This instrument measures the total heat flow Φ_R due to the radiation falling on an element of a plane surface.

Principles of the instrument and their realization

All the radiation falling on the small circular orifice O (Fig. 3.2) is focused by the ellipsoidal mirror onto the thermopile which gives a potential difference as a linear function of the energy received

(i) the orifice O is thin-rimmed so as to allow all the radiation to enter without parasitic reflection, whatever the angle of incidence (Fig. 3.3)

(ii) the thermopile consists of a hemispherical receiving pellet, a cylinder AB and a cooled metal mass M, the three parts being combined in a common stainless steel block (25% nickel, 25% chromium), in order to obtain a regular flow of heat. The heat sensitive pellet P, absorbs practically all the radiation (from 95% to 98% whatever the wavelength) intercepted by its blackened and oxidized surface, *i.e.* all the radiation entering at O, as we shall see later. (Under steady-state conditions, the rate of heat removal by the cooled metal mass M is equal to the rate of heat collection by the pellet.) Constantan wires soldered into A (hot soldering) and into B (cold soldering), together with the 20–25 stainless steel cylinder AB, form a thermocouple which produces an e.m.f. proportional to the energy received by P, at least so long as the temperature of the pellet does not rise too far, *i.e.* while the losses due to radiation and convection are negligible. In view of this it is obvious that the thermopile zero corresponds to the energy radiated by a black body enclosure raised to the temperature of the instrument, which is effectively zero, as can be seen from Fig. 3.4, when this temperature is that of the surroundings. The thermopile is proportioned so that its equilibrium temperature remains low under ordinary operating conditions;

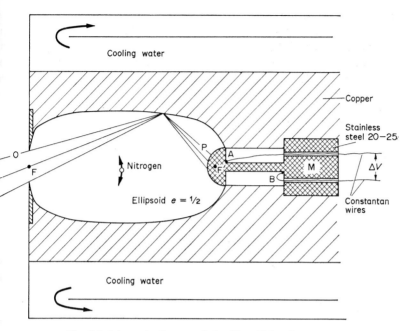

Fig. 3.2 Schematic diagram of the ellipsoidal radiometer

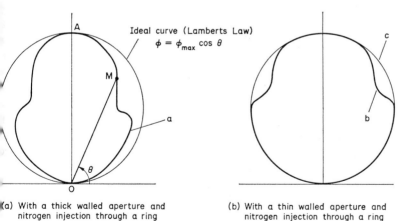

(a) With a thick walled aperture and nitrogen injection through a ring

(b) With a thin walled aperture and nitrogen injection through a ring

(c) With a thin walled aperture and nitrogen injection at two points

Fig. 3.3 Influence of incidence on the transmission factor of the instrument

(iii) the orifice O and the receiving pellet P are centred on the foci F and F
for which the ellipsoidal mirror is stigmatic. Dispersion due to the non
zero diameter of the orifice O does not give rise to difficulties because
of the hemispherical shape of P, the diameter of which is slightly larger
than that of O. All radiation is received after two reflections at the most
from the mirror;

(iv) the ellipsoidal mirror with an eccentricity of $\frac{1}{2}$ is made of coppe
covered with a thin layer (0·05 mm) of polished gold, and has a reflection

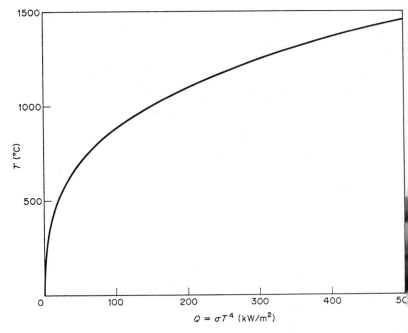

Fig. 3.4 Radiation from the black body furnace

coefficient very close to unity which does not depend either on the angle
of incidence or on the wavelength up to 2μ. We lay great stress on the
necessity for accurate workmanship in the mirror in order to avoid
diffusion of the radiation;

(v) a current of dry nitrogen injected through small holes (from 2 to 8
distributed in the plane of the minor axes of the ellipsoid prevents the
entry of combustion gases and particles into the device. In this way we
avoid condensation and deposits of solid and liquid particles on the
mirror. The dimensions and number of the holes for admitting nitrogen
must be small enough not to affect the efficiency of the mirror (Fig. 3.3

Calibration

The ellipsoidal heat flow meter, given the characteristics of its thermopile, does not provide an absolute indication and has to be calibrated by subjecting it to the radiation of different black bodies of known temperature (Fig. 3.5). Experience shows, as we have already pointed out, that in the normal range of operation (0 to 500 kW/m²), the millivolts at ambient temperature being negligible, the calibration curve passes through the origin so that its determination can be made with one black body only.

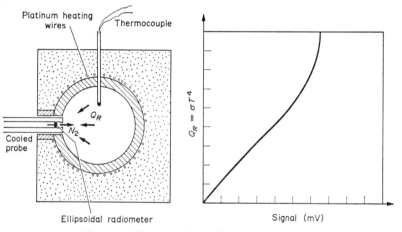

Fig. 3.5 Calibration of the ellipsoidal radiometer

Calibration should be carried out frequently during the measurements for several reasons:

i) despite the injection of nitrogen, a few particles or droplets of oil may manage to get into the ellipsoid and settle on the mirror (generally in the region of the orifice or the pellet) which in this way becomes selective for certain incidences and loses its efficiency;

ii) as a result of mechanical shock and chance heating the position of the thermopile may be disturbed and its receiving surface damaged;

iii) the area of the orifice O may change.

It should be noted that the nitrogen injected into the ellipsoid in order to protect the mirror produces convection currents which cool the receiving pellet and modify the signal, as shown in Fig. 3.6. It is essential, therefore, to control the flow of nitrogen and to keep it constant during the whole of the measurements and calibration. In the case of the apparatus from which Fig. 3.6 was derived, the operating flow was taken as 45 l/h, a value which can be varied slightly without involving any modification of the sensitivity.

Finally, it should be added that during calibration the cooled head o
the probe, and above all the current of nitrogen, lower the temperature o
the internal surface of the black body. We suggest that the calibratio
should be carried out quickly as follows: wait for equilibrium to b
established without nitrogen, inject the gas only during the time needed t

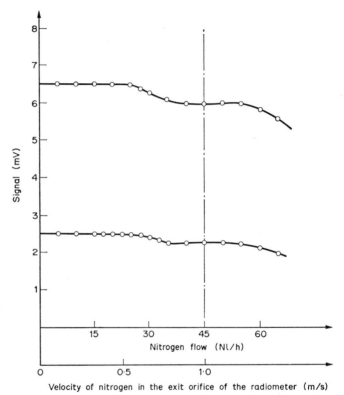

Fig. 3.6 Influence of the nitrogen flow rate on the signal of the ellipsoidal radiometer

obtain the required value and then check to see that the origina
equilibrium is restored after cutting off the nitrogen.

Performance

The response time of the ellipsoidal heat flow meter is of the order of on
minute and does not, therefore, allow rapid changes in the flame radiatio
to be followed (it would be necessary for this to use a thermopile of lowe
inertia like that of the total radiation pyrometer, see Section 4.1). When th

constant does not change we have been able to obtain very good reproducibility of the measurements (the small percentage variation is due
largely to fluctuations in the flame).

The accuracy of these measurements depends, therefore, on the care
taken during calibration in the first place (measurement of the black-body
temperature, positioning of the probe in the sighting hole, etc.) on the
control of the nitrogen flow, and finally on the more-or-less difficult
conditions of measurement.

Normally the heat flow received by radiation can be measured to better
than 5%. Finally, the maintenance of the equipment can be reduced to
periodic inspections for cleaning the mirror, blackening the pellet P, and
checking the position of the thermopile.

3.3 Conductivity plug-type heat flow meter

This instrument measures the total heat flow (convection plus radiation)
absorbed by its receiving surface raised to a known temperature.

Principles and realization

Under steady-state conditions the heat flow Φ transmitted by conduction
in one direction is given by

$$\Phi = -\lambda \, \mathrm{d}t/\mathrm{d}x \tag{3.2}$$

Where λ is the thermal conductivity, a physical property of the material
    used, which is a function of the temperature t, and
$\mathrm{d}t/\mathrm{d}x$ is the temperature gradient in the direction of flow to the right
    of section A.

If the transmission takes place only in the axial direction in a cylinder (of
constant section A) as shown in Fig. 3.7, a knowledge of the law $\lambda(t)$ enables us to determine the flow Φ from the temperature measurements at
two known levels x_1 and x_2. Integration of equation (3.2) gives:

$$\Phi \int_{x_1}^{x_2} \mathrm{d}x = - \int_{t_1}^{t_2} \lambda \, \mathrm{d}t$$

$$\Phi = - \frac{1}{x_2 - x_1} \int_{t_1}^{t_2} \lambda \, \mathrm{d}t$$

If the mean value of λ_m between the temperatures t_1 and t_2 is defined by
the equation

$$\lambda_m = \frac{\displaystyle\int_{t_2}^{t_2} \lambda \, . \, \mathrm{d}t}{t_2 - t_1}$$

we can find the flow from

$$Q = - A\lambda_m \frac{(t_2 - t_1)}{(x_2 - x_1)}$$

The curve (Fig. 3.8),

$$y(t) = \int_0^t \lambda(t)\, dt$$

can be used to give the value of λ_m since

$$\lambda_m = \frac{\int_{t_1}^{t_2} \lambda \cdot dt}{t_2 - t_1} = \frac{\int_0^{t_2} \lambda \cdot dt - \int_0^{t_1} \lambda dt}{t_2 - t_1}$$

This is the principle of the conductivity heat flow meter, consisting, as shown in Fig. 3.7, of a cylinder protected by concentric guard rings which ensure the axial flow of the total heat Q_T (radiation and convection) absorbed by the receiving surface to the other end, which is cooled by water. The nature of the material forming the cylinder is chosen for a given length of the instrument as a function of the required surface temperature: copper for low temperatures, stainless steel for medium, and alumina for high temperatures. There are two thermocouples (thermocoaxial) of nickel-nickel chromium, inserted as far as the axis of the cylinder through holes 1 mm in diameter which measure the temperatures t_1 and t_2, the level of 1 being close to the receiving surface, and that of 2 several millimeters above the plane where the guard rings are joined. It is assumed that the radial heat exchange is negligible when the device comprises at least two guard rings separated by a very thin layer of air. We can then express Φ_T by

$$\Phi_T = \varepsilon_s \Phi_R - \varepsilon_s \sigma T_s^4 + \Phi_c = - \frac{y(t_2) - y(t_1)}{x_2 - x_1}$$

where: ε_s = emissivity of the receiving surface

T_s = temperature of the receiving surface

Φ_R = heat flow received by radiation at the receiving surface and measured by the ellipsoidal heat flow meter

$\Phi_c = h(T_G - T_s)$, the heat flow received by convection by the receiving surface, T_G being the temperature of the gas above the instrument

It is possible, therefore, to calculate Φ_T so long as we know $\lambda(t)$, *i.e.* $y(t)$, T_s and ε_s.

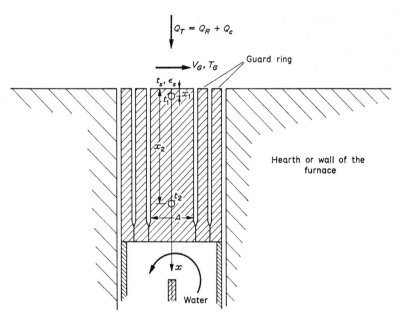

Fig. 3.7 Schematic diagram of the principle of the conduction heat flow meter

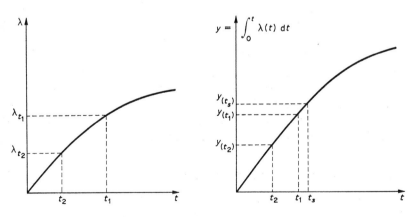

Fig. 3.8 Variation of the thermal conductivity (λ) with temperature (t)

Determination of the thermal conductivity coefficient

There are, in fact, three possibilities:
(i) to call on a laboratory specializing in making this measurement (accuracy of $\pm 1\%$);
(ii) to use the mean values given in Tables (accuracy of $\pm 10\%$ for stainless steel, less good for copper because of the marked effect of impurities);
(iii) to carry out the measurement by subjecting the instrument with its receiving face blackened beforehand either by soot or a special paint to the radiation of a black body furnace; the heat flow absorbed by the receiving surface is that emitted by the black body:

$$\Phi = \varepsilon_{sn}\sigma(T_F^4 - T_{sn}^4) + h_c(T_F - T_{sn})^{1\cdot 25} = -\frac{t_2 - t_1}{x_2 - x_1}\lambda_m$$

hence

$$\lambda_m = \frac{x_2 - x_1}{t_1 - t_2}[\varepsilon_{sn}\sigma(T_F^4 - T_{sn}^4) + h_c(T_F - T_{sn})^{1\cdot 25}]$$

where
T_F is the temperature of the black body furnace,
ε_{sn} is the coefficient of emissivity of the blackened surface, estimated at $0\cdot 98$,
T_{sn} is the temperature of the blackened surface obtained by extrapolation of the temperatures t_1 and t_2, assuming that λ is independent of t, and
$h_c(T_F - T_{sn})^{1\cdot 25}$ is a term giving the quantity of heat received by natural convection in the furnace. It has been found (5) that for our arrangement $h_c \simeq 4\cdot 7$ Kcal m²h°C$^{1\cdot 25}$

The precision obtained depends on the care given to the measurements, but it seems that it could hardly be better than $\pm 5\%$. In any case we only obtain the mean value of λ applicable between the temperatures t_1 and t_2.

In the course of the measurements one must, of course, take care that the temperature of the metal does not reach values which are high enough to cause any modifications of the structure and therefore of the conductivity. This is why it is desirable to make periodic checks on the value of λ.

Determination of the surface temperature

We have seen that thermocouple 1 was placed near the receiving surface ($x_1 = 2$ mm). Hence a small error is involved in calculating t_s by linear extrapolation of the temperatures t_1 and t_2:

$$t_s = t_1 + x_1\left(\frac{t_1 - t_2}{x_2 - x_1}\right)$$

A more exact estimate of t_s is obtained from the curve $y(t)$:

$$\frac{y(t_s) - y(t_1)}{x_1} = \frac{y(t_1) - y(t_2)}{x_2 - x_1}$$

$$y(t_s) = x_1 \left(\frac{y(t_1) - y(t_2)}{x_2 - x_1}\right) + y(t_1)$$

from which t_s is obtained by relating $y(t_s)$ to the curve $y(t)$.

The error in the surface temperature determined in this way does not exceed a few degrees Celsius, on the condition, of course, that the radial heat flow remains negligible, which is not always true so close to the receiving surface.

Determination of the surface emissivity

Before the instrument is used for measurements its surface is oxidized by raising it to a high temperature for several hours in an oxidizing atmosphere. The surface emissivity of stainless steel is then about 0·85 and varies very little with temperature. It may change, however, as a result of soot or ash deposition, and it is advisable to re-check from time to time by subjecting the instrument to black body radiation

$$\varepsilon_s = -\frac{1}{T_F^4 - T_s^4}\left[\frac{y(t_2) - y(t_1)}{x_2 - x_1} + h_c(T_F - T_s)^{1\cdot25}\right]$$

Actually, when the surface is covered with soot the latter burns off in the air during heating. One can then try measuring in a neutral atmosphere, or simply take the value $\varepsilon = 0\cdot98$ without risking any serious error. The resulting accuracy for ε_s is of the order of 5%.

Performance

The response time of a conductivity plug-type heat flow meter depends on its dimensions and on the nature of the material used in it. In the case of the stainless steel devices, this time is of the order of 10 minutes. We have already said that the surface temperature was measured to a high degree of accuracy (within a few degrees) and that the emissivity of the same surface is known to within 5%. As for the total heat flow received, this is measured with an accuracy very close to that of λ, that is, better than 2% if the determination of λ has been made by a specialized laboratory. Finally, determination of the convection, when the radiation is measured to within 5% by the ellipsoidal heat flow meter is possible with a satisfactory absolute precision ($\pm25\ \text{kW/m}^2$) in the majority of cases encountered at the Foundation.

3.4 Circulating water heat flow meter

This apparatus, shown in Fig. 3.9, measures the total heat Φ_T received by a cold surface, s, $(t < 80°C)$, which therefore radiates negligible energy,

$$\Phi_T = \varepsilon_s \Phi_R + \Phi_c$$

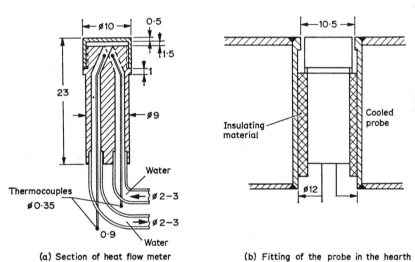

(a) Section of heat flow meter　　　(b) Fitting of the probe in the hearth

Fig. 3.9 The circulating water heat flux meter

The receiving surface is a metal pellet cooled by a current of water of which the flow q, the input temperature t_1, and the outlet temperature t_2 are measured:

$$\Phi_T = q(t_2 - t_1)c$$

where c is the specific heat of water.

This pellet, encased in a protective probe, must be thin enough for the surface temperature T_s to remain close to t_2. Its emissivity is measured by exposure to a black body furnace. The response time is relatively short (10 seconds) but the accuracy is rather poor due to possible lateral losses, the difficulty in maintaining a constant small flow of water, and the error in the measurements of the inlet and outlet temperatures by thermocouples.

3.5 The Onera heat flow meter

Like the conductivity heat meter this device measures the total heat flow absorbed by the receiving surface, the temperature of which can be raised to different levels.

This meter is illustrated by the sketch in Fig. 3.10; a thin pellet (1 mm to 2 mm) is mounted and centred on three points in a prepared cavity in a body C heated by electrical resistances R. Two thermocouples, one soldered to the centre of the pellet P, and the other to the wall of the cavity in the body C, give the temperature of the receiving pellet T_S and that of the body T_C respectively. The measurements are carried out as follows: the body C is heated outside the furnace to the temperature T_C at which we wish to operate, the pellet being kept cooler ($T_s < T_c$). The equipment

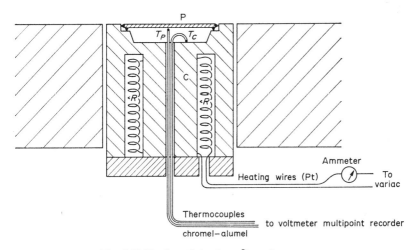

Fig. 3.10 Section of the Onera heat flow meter

is then introduced at the measuring point where the pellet P receives the heat flow to be measured Φ_T and is heated in such a way that T_S finally reaches and passes T_C. At the moment when $T_C = T_S$ the exchange of heat between the pellet P and the body C is zero, and the law of heating a pellet of surface S, mass m and heat capacity C is then

$$\Phi_T = \frac{mC}{S}\, dT_S/d\theta$$

where $dT_s/d\theta$ is the increase in the temperature T_S with respect to time. The value of $dT_S/d\theta$ is found graphically from the recorded temperatures T_S and T_C carried out by means of a two-way electronic potentiometer (Fig. 3.10) and the quantity mC/S is given as a function of temperature by a curve provided by a laboratory specializing in the determination of heat capacities.

By modifying the electric heating of the body C, one can repeat the measurement up to the temperature level which is judged to be necessary (Fig. 3.10) for plotting the curve $\Phi_T(T_s)$ of Fig. 3.1:

$$\frac{mC}{S} \cdot \frac{dT_s}{d\theta} + \varepsilon_s \sigma T_s^4 = -h_c T_s + (h_c T_G + \varepsilon_s \Phi_R)$$

and for determining h_c and Φ_R.

The emissivity ε_s of the receiving pellet should be measured again by subjecting the device to black body radiation.

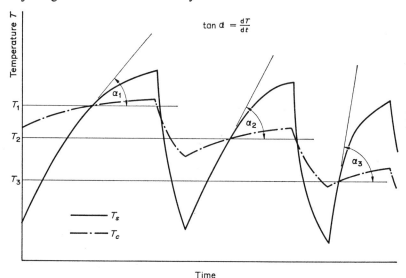

Fig. 3.11 Temperature recording T_s and T_c of the heat flux meter

The body C and the receiving pellet P can be made of different materials according to the maximum operating temperatures: inconel, platinum or aluminium. The receiving face of the pellet should be oxidized (at low temperature blackened with a special paint (4)) and the other face polished in such a way as to restrict the exchange by radiation with the body C.

This heat flow meter which has given satisfactory results in the laboratory (6) is not suitable for the conditions of special measurements (7)(8)(9). Actually the Foundation only uses conductivity and ellipsoidal total radiation types of heat flow meters, either mounted on the end of a cooled probe for determining the heat exchange at the vertical walls, Figs. 3.12 and 3.13, or laterally on probes of rectangular section for measurements at furnace level (2). In the latter case, two devices are mounted on the same probe in order to carry out the necessary two measurements simultaneously (Fig. 3.14).

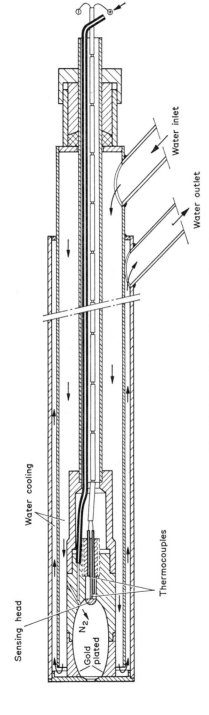

Fig. 3.12 Normal ellipsoidal radiometer probe

Water inlet

Water outlet

Water cooling

Thermocouples

Sensing head

N_2

Gold plated

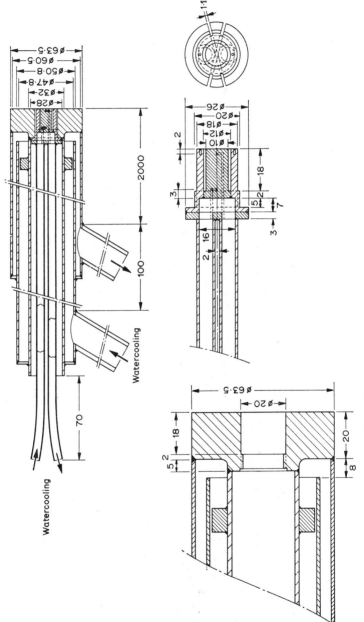

Fig. 3.13 Conductivity plug-type heat flow meter

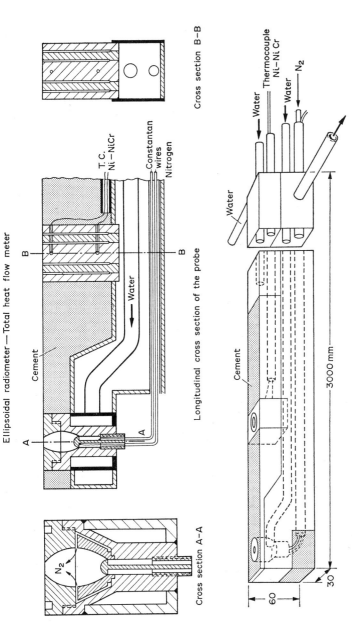

Ellipsoidal radiometer—Total heat flow meter

Cross section B–B

T. C.
Ni –NiCr

Constantan
wires

Nitrogen

B

B

Water

Cement

A

A

Longitudinal cross section of the probe

Cross section A–A

N₂

Water

Thermocouple
Ni –Ni Cr

Water

N₂

Water

Cement

3000 mm

60

30

Fig. 3.14 Probe to measure heat transfer

Chapter 4

Measurement of Flame Radiation

This is a question of determining the emissive power of a particular flame thickness or of the complete flame. The measurement is carried out with a total radiation pyrometer of BISRA-KNHS type, an instrument which gives the intensity of the radiation received in respect of a specified direction.

4.1 Total radiation pyrometer

The pyrometer comprises a spherical concave mirror with an aluminium reflecting layer, protected by a thin coating of silicon; this mirror focuses the radiation received in a very small solid angle onto the receiving pellet of a thermoelectric cell (Fig. 4.1).

Practical realization

Although this pyrometer is extremely simple in principle it is advisable to take certain precautions in designing it.

(i) The receiving angle is controlled either by the interior diameter and the length of the sighting tube, or by a system of screens arranged in front of the thermopile, or by both methods at the same time. The sighting tube must be blackened inside and fitted with diaphragms for protection against parasitic reflections. If this tube is very long and the supports which hold it are flexible it will adopt sufficient curvature under its own weight to cause a displacement of the image on the thermopile. This displacement becomes apparent through a reduction in sensitivity resulting from the diminution of the surface intercepted by the image.

(ii) The mirror is pierced at its centre with a sighting hole which makes it possible to control simultaneously the aiming of the device, the state of the sighting tube and the position of the image on the thermopile. The makers of the mirror (Balzers, Liechtenstein) use a hole of 9 mm minimum

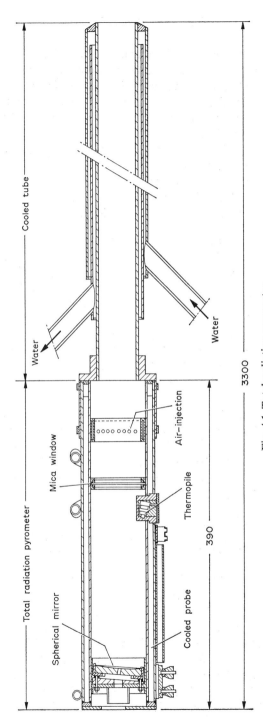

Fig. 4.1 Total radiation pyrometer

diameter, which implies that the diameter of the sighting tube must be definitely greater than this value. A more precise measurement procedure (the smallest tube that we have made is 17 mm) requires a modification of the mirror. The direction of the mirror can be adjusted by rotation about two perpendicular axes. Normally the axes of the mirror and the sighting tube make an angle of 5°.

(iii) Various thermopiles have been used, the first being a Kipp thermopile with a sensitive element 9 mm in diameter. New high sensitivity types have now appeared on the market. Their use, which allows the sighting tube to be reduced, *i.e.* an improvement in the localization of the measurement, involves a certain number of precautions:

(*a*) the thermopile must be placed in an enclosure of considerable thermal inertia, with a temperature which is uniform, and if possible constant in time. It is necessary to prevent any convection movement in this enclosure and to avoid heating the thermopile too much;

(*b*) if there are likely to be even small variations of temperature with time it is advisable to use a compensated thermopile (with two cells— an environmental cell in opposition to a measurement cell) in such a way that it reverts to zero electrically in the absence of radiation. Another possibility consists of placing a shutter between the mirror and the thermopile in order to check the zero between each measurement;

(*c*) if the image is smaller than the receiving cell the signal may depend on its position on this cell, so that the best arrangement is when the image covers the thermopile exactly. Unfortunately, a slight displacement of the image in this case due to a lack of rigidity in the sighting-tube + mirror + thermopile system causes a marked reduction in sensitivity.

(iv) Finally, in order that the reading shall be good there must not be any absorbing screen or emitter between the thermopile and the body (flame or solid) being studied. For this it is necessary:

(*a*) to inject air into the sighting tube in order to clear it of absorbing gases (H_2O, CO_2, etc.) as well as dust in suspension, which is liable to enter, particularly when the furnace is under pressure. The measurements will, therefore, be made as a calibration under the hygrometric conditions at the moment, variations of which are of little importance because of the low effective partial pressures of water vapour;

(*b*) to refrain as far as possible from placing windows and lenses in the trajectory of the radiation so as to avoid selective absorptions as a function of wave length. If this is not possible, it is necessary to make corrections to the reading, corrections which are possible only if we assume a definite distribution of energy as a function of the wave length and the temperature. We should mention, however, that certain materials have a constant absorption coefficient in the infra-red.

Calibration—performance

In order to calibrate the device we subject it to radiation from a black body raised to different temperatures. The curve giving the signal as a function of the energy received ($\Phi = \sigma T^4$) is a straight line so long as the thermopile is not heated too much.This calibration must be checked frequently in order to bring out any possible variations due to disturbances in the optical system, dust settling on the screens and the mirror, or to a partial obstruction of the sighting tube. The slope of the line obviously depends on the sensitivity of the thermopile and on the diameter and aperture of the intercepted beam.

The response time, which is that of the thermopile, is short enough to follow individual fluctuations in the flame (0·01 second for air thermopiles, 0·1 second for vacuum thermopiles).

Probe

The probe used in our laboratories is illustrated in Fig. 4.1. This comprises the pyrometer proper fixed in a cooled sighting tube, intended to be inserted into the furnace for carrying out the radiation measurements described below.

4.2 Schmidt method

While the use of total radiation pyrometers, which are extensively employed in industry, is well known so far as the measurement of surface temperatures is concerned, their application to the determination of flame emissivity and temperature is not so well known.

Schmidt was the first to suggest a method of calculating the total emissivity factor of a flame on the basis of three radiation intensity measurements, using the total radiation pyrometer.

R_1 = radiation intensity of the flame alone, obtained by viewing a cold target through the flame, $= \varepsilon\sigma T_F{}^4$

R_2 = radiation intensity measured by viewing a hot black body at a temperature T_{CN} through the flame, $= \varepsilon\sigma T_F{}^4 + \tau\sigma T_{CN}{}^4$

R_3 = radiation intensity due to the black body calculated from its temperature $T_{CN} = \sigma T_{CN}{}^4$.

We can in fact, write

$R_2 = R_1 + \tau R_3$;

  τ is the over-all transmission coefficient of the flame,

  ε is the over-all emission factor for the flame,

T_F is the temperature of the flame, which is assumed to be constant
  throughout its thickness.

$$\tau = 1 - \alpha - r$$

where r is the coefficient of reflection or diffusion.

If the flame behaves like a grey body its absorptivity and emissivity
are equal, and if we assume that the scattering is zero ($r = 0$) we have:

$$\varepsilon = 1 - \tau = 1 - \frac{R_2 - R_1}{R_3}$$

$$T_F = \left[\frac{R_1 R_3}{\sigma(R_3 + R_1 - R_2)} \right]^{\frac{1}{4}}$$

In this way we can calculate the emissivity and the temperature of the
flame on the basis of the three measurements R_1, R_2 and R_3.

It should be remembered that Schmidt has proposed his method for
flames which behave like a grey body, are not scattering, and which have a
uniform temperature throughout the thickness under consideration. In
the absence of scattering the method is always exact monochromatically
(which reduces its importance considerably) and in flames which contain
non-grey gases (CO_2, water vapour, CO, etc.), so long as the black body
which is used for R_2 and R_3 is at the same temperature as the gases.
Finally, Schmidt's method is never in practice strictly applicable to industrial diffusion flames in which the temperatures and the concentrations
of radiating particles and gases sometimes vary considerably. However, its
results are acceptable in the case of fuel oil flames which radiate mainly
by the soot, when the temperature of the flame and that of the black body
furnace are close to each other (**11**). Apart from these special cases the
calculated values of ε and T_F do not have an exact physical meaning.

It should also be added that in practice one does not use real black
bodies for determining R_2 and R_3 but simply a section of refractory lining
of low conductivity and considerable thickness. The energy balance
shows that, in the absence of convection heat exchange and losses due to
conduction on the lining, the apparent emissivity of the surface is 1;

actually, if Φ_T is the total energy given out by the lining and T_S is the surface temperature measured by a thermocouple,

$$\Phi_T = \varepsilon_s \sigma T_s{}^4 + (1 - \varepsilon_s)\Phi_i$$

where Φ_i is the total incident energy which is reflected in the proportion of $(1 - \varepsilon_s)$. As there are no heat exchanges apart from those due to radiation, the rate of emission must equal the rate of absorption, and therefore:

$$\sigma\varepsilon_s T_s{}^4 = \varepsilon_s\Phi_i$$

that is

$$\Phi_i = \sigma T_s{}^4$$

therefore

$$\Phi_T = \varepsilon_s \sigma T_s{}^4 + (1 - \varepsilon_s)\Phi_T$$

i.e.

$$\Phi_T = \sigma T_s{}^4$$

which shows clearly that the apparent emissivity is 1 whatever the actual emissivity may be.

Mayorcas and Riviere have verified that the walls of the IJmuiden furnace have an effective apparent emissivity equal to unity, which enables us to avoid the difficult handling of black body furnaces. It should be pointed out, however, that if the apparent emissivity is that of a black body the energy distribution may be quite different from that given by Planck's Law and this will be all the more marked the less the actual emissivity of the refractory assumed to be grey. A further error may result in the case of non-grey flames.

Finally we should mention that the firm Land Pyrometer Limited of Sheffield (England) have marketed a pyrometer which enables R_1 and R_2 to be read simultaneously (twin beam radiation pyrometer) **(12)**.

4.3 Measurement of local radiation

The measurements carried out on the experimental flames in our furnaces give, as we have seen in the preceding Sections, both the local temperature of the gas and the local concentrations of particles (with the granulometric distribution) and of gases. It is also equally important to measure a quantity which characterizes the radiation in order to establish the physical laws which relate this to the other variables on which it depends; this quantity is the local extinction coefficient δ which is defined by the following ratio (assuming the emission is equal to the absorption)

$$\delta = \frac{\dfrac{d\Phi_R}{dy}}{\dfrac{\sigma}{\pi} T_G{}^4 - \Phi_R(y)}$$

where y is the thickness of the flame and Φ_R is the flux radiated in the direction of measurement (in kW/m²).

In the special case when T_G is constant, integration of the above equation leads to Beer's Law.

$$\delta = -\frac{1}{y}\log\left(1 - \frac{\Phi_R(y)}{\frac{\sigma}{\pi}T_G{}^4}\right)$$

or

$$\Phi_R(y) = \frac{\sigma}{\pi}T_G{}^4(1 - e^{-\delta y})$$

There are, in fact, two methods of measuring δ.

Integration of radiation

This is the "traversing method" of Beér and Claus.

It consists in plotting the curve $\Phi_R(y)$ experimentally while viewing a cold target mounted on the furnace wall, by means of a total radiation

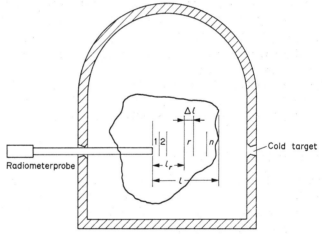

Fig. 4.2 Traversing method of radiation measurement

pyrometer with an extended sighting tube, through a flame thickness varying from $y = 0$ to $y = L$, L being the width of the furnace (Figs. 4.2 and 4.3).

We can read $\Phi_R(y)$ directly from this curve and then evaluate $d\Phi_R(y)/dy$ by drawing the tangent to the abscissa y. These data enable us to calculate δ with rather poor accuracy, mainly because of the error in drawing the tangent.

In any case, Beér and Claus have calculated by integration the curves of $\Phi_R(y)$ **(11)** for an oil flame by making the following assumptions:

(i) the flame radiation is grey;
(ii) the gas radiation is negligible compared with that of the soot;

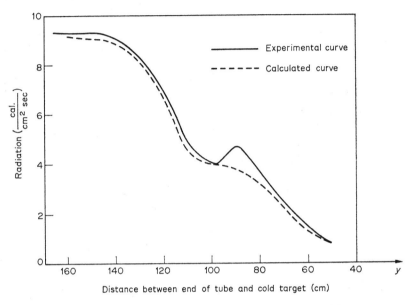

Fig. 4.3 Development of radiation as a function of flame thickness

(iii) there is no scattering of radiation due to soot particles. This assumption is justified on the grounds of the small size of the soot particles (this is not the case for pulverized coal flames);
(iv) the temperature of the flame is constant over the thickness under consideration.

Under these conditions it can be shown that δ is proportional to the soot concentration and to the absolute temperature T_G, and it has been confirmed that the calculated curves are in good agreement with those derived experimentally. This indicates that the assumptions are justified and that as a result Schmidt's method is applicable to oil and gas flames.

In the same way, Hemsath **(13)** has put forward a method for calculating the radiation from pulverized coal flames, given the distribution of the gas and solid concentrations, the specific surface of the particles, and the temperature. An exhaustive study of this problem has also been made by Gouffe **(14)**.

Local suction radiometer

In order to measure the local radiation of the flame directly and to calculate the local extinction coefficient, a new radiometer is in the process of development.

As Fig. 4.4 shows, this instrument consists of two concentric tubes of refractory material, with four holes half way down which are intended for drawing in gas from the flame towards two outlets arranged at the ends.

In this way we have a column of gas which is associated with a definite point in the flame, the length of which is limited by the injection of blast air at each end of the inner tube. This column of hot gas is viewed by a standard radiation pyrometer (KNHS–BISRA type) on one side, and closed off by a cold, non-reflecting screen on the other. Hence the total radiation pyrometer only receives the radiation due to the column of gas which has been drawn in.

We have already seen that:

$$\delta = -\frac{1}{y}\log\left(1 - \Phi_R(y)\bigg/\frac{\sigma T_G^{\,4}}{\pi}\right)$$

that is, the suction radiometer enables the local extinction coefficient to be determined from the above equation simply by knowing the gas temperature T_G and by measuring the radiation $\Phi_R(y)$ due to a column of gas length $y = L$.

The object of the refractory screening tubes (as we can see in Fig. 4.4, the diameter of the internal tube is greater than that of the total radiation pyrometer viewing tube) is not only to define the column of hot gas, but also to limit the exchange of heat between this gas, the flame and the enclosure.

This instrument was used for the first time during the tests 0–17 in the Spring of 1967, in order to compare its measurements with those of the total radiation pyrometer when viewing a cold target through the flame.

Figure 4.5 (a) illustrates the character of the radiation through an oil flame measured with the two instruments described above, and Fig. 4.5 (b) gives the temperature distribution and the extinction coefficient. It can be confirmed that the two methods agree as far as oil flames are concerned.

4.4 Black body calibration furnaces

As we have already seen above, all radiation and heat transfer measuring devices must be standardized against a source which supplies a known energy by radiation alone. Such a source is a black body furnace **(15)**.

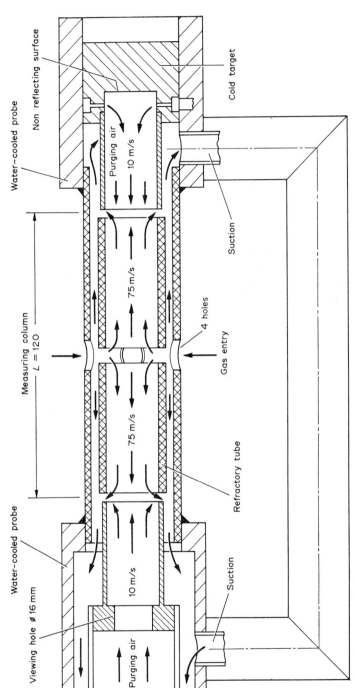

Fig. 4.4 Diagram of suction radiometer (prototype for 0–17 trials)

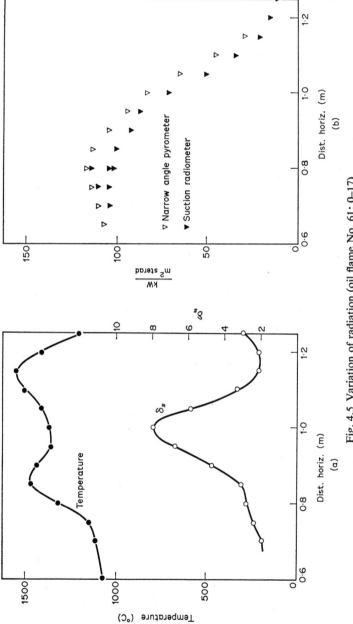

Fig. 4.5 Variation of radiation (oil flame No. 61; 0-17)

10 p

Combustion – measurement

Chedaille, J.

Measurements in flames Edward
Arnold 1972 xi 228 p

Ref. QD 516 . I 52 (Bsт)

29434375

Principle and effective emissivity

We recall briefly that, by definition, a black body is one which absorbs completely the incident radiation, and emits by radiation an energy which is a function of its temperature alone, in accordance with the Stefan–Boltzmann Law.

$$\Phi_T = \sigma T^4 \quad \text{where } \sigma = 5 \cdot 67 \times 10^{-5} \text{ erg/cm}^2 \text{ s (K)}^4$$

In addition the distribution of the emitted radiation as a function of the wavelength λ is given by Planck's Law:

$$d\Phi_T/d\lambda = C_1 \lambda^{-5} \left/ \exp\left(\frac{C_2}{\lambda T}\right) - 1 \right.$$

where
$$C_1 = 3 \cdot 7403 \times 10^{-5} \text{ erg cm}^2/\text{s}$$
$$C_2 = 1 \cdot 4387 \text{ cm K}$$

Finally, a body is said to be grey if it emits (or absorbs) a constant fraction of the corresponding black body radiation; this is the emissivity of a grey body, an emissivity which is equal to the absorptivity.

In practice a black surface is made by making an opening in a closed grey enclosure which has a uniform temperature, and an internal area which is large compared with that of the opening. Thus any radiation which is incident on the opening, has practically no chance of coming out again except after numerous reflections, that is, after almost total extinction, which is a function of the absorption coefficient of the internal surface.

It is understood, of course, that the effective emissivity ε_0 of a black body furnace of this kind depends on both the geometry and the actual emissivity ε of the internal surface. Gouffe has calculated ε_0 theoretically by taking into account all possible internal reflections:

$$\varepsilon_0 = \frac{1 \ (1 - \varepsilon)(s/S - \sin 2\theta)}{1 + s/S \left(\dfrac{1 - \varepsilon}{\varepsilon}\right)}$$

where s is the area of the opening;

S is the internal area of the furnace;

2θ is the angle at which the opening is viewed from the back of the furnace.

$$tg 2\theta = 2R/L$$

where R is the radius of the opening which is assumed to be circular, and L is the distance between the opening and the back of the furnace.

Michaud (**16**), by examining certain simple geometrical shapes

$$\left(\text{spheres } s/S = \frac{1}{2 + \left(\dfrac{L}{R}\right)^2}, \text{ cylinder } s/S = \frac{1}{s\left(1 + \dfrac{L}{R}\right)},\right.$$

$$\left.\text{cone } s/S = \frac{1}{1 + \sqrt{1 + \left(\dfrac{L}{R}\right)^2}}\right)$$

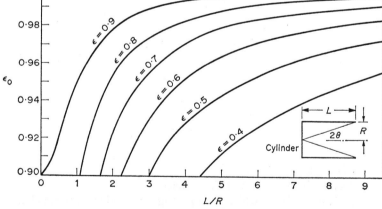

Fig. 4.6 Emissivity of black body furnaces

has verified Couffe's relationship experimentally: Fig. 4.6 shows the results obtained for the two shapes which are of main interest—the sphere and the cylinder—for different values of the ratio L/R and of the emissivity.

Realization

This is a question of determining the most convenient shape to make, knowing the radius R which is fixed by the size of the probes, while retaining a satisfactory emissivity. As a rule the spherical form is preferred to the cylindrical because uniformity of temperature is easier to obtain, at least so long as the heating elements can be wound round the sphere.

This can be done up to temperatures below 1100°C, using Kanthal or nickel chromium elements, and up to temperatures of the order of 1250°C with Kanthal A_1 or pure platinum elements. In the case of higher temperatures it is best to use carbon resistance elements, and then the cylindrical furnace design is more convenient.

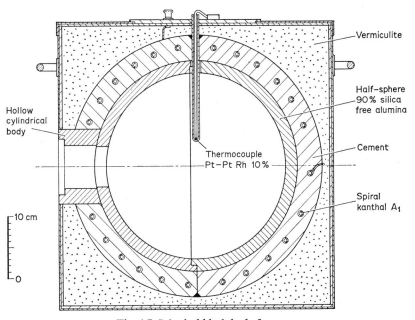

Fig. 4.7 Spherical black body furnace

Spherical furnace. The furnace built in our laboratories at IJmuiden is shown in Fig. 4.7. This comprises:

(i) an internal sphere made of coarse-grained refractory material containing 90% alumina without silica, which has an emissivity between 0·5 and 0·7 at high temperature (1200°C) so that the effective emissivity ε_0 of the furnace with $R = 35$ mm and $L = 300$ mm ($L/R = 8\cdot5$) is close to 0·99. This 2 cm thick sphere was bought from the Société des Electrodes et des Réfractaires SAVOY (Paris);

4

(ii) heating elements rolled in a spiral about their own axis and over the internal sphere, and embedded in a layer of special cement (Sillimanite Ramming Mix), 5 cm thick;

(iii) the overall external diameter of the furnace is 44 cm, with an internal diameter of 30 cm. It is placed in alumina powder held in a metal container with the opening extended by a protective refractory cylinder, 65 mm long. A Pt–PtRh10% thermocouple protected by a pure calcined alumina sheath, which is inserted about 10 cm into the sphere, serves to measure the equilibrium temperature.

For the purpose of construction the furnace is made in two hemispherical halves which are then assembled and bonded with a special cement. Practical details concerning these furnaces have been given by R. R. Kissel (15). We give the more important of these below:

(i) each hemisphere is heated by an element 33 m long and 1·6 mm diameter in the form of a spiral 5 mm in diameter and 8 m long. The resistance of each element of Kanthal A_1 is approximately 24 Ω (this varies very little with temperature);

(ii) a Variac is used to adjust the supply to the two elements in series. The maximum consumption (1250°C) is about 5 kW at 220 volts;

(iii) the first heating and cooling cycle for the furnace must be carried out very slowly (45°C per hour up to 700°, and then 25°C per hour up to 1200°C, when heating, and 20°C per hour when cooling) in order to allow the heating element to take up any slack and to withstand better more rapid changes of temperature (up to 30°C per hour) because of the consecutive differential expansions;

(iv) in order to prolong the service life of the furnace it is advisable to keep to the same heating and above all to avoid frequent reduction of the temperature below that of re-crystallization of the metal (about 460°C). This reason, together with the considerable thermal inertia of the furnace, makes it necessary to have as many black body furnaces available as there are energy levels likely to be required. When the furnaces are not used for periods up to two months they should be kept at 600°C (0·2 kW consumption) rather than shut down completely.

Cylindrical furnace. We have also used a cylindrical furnace which had graphite heating elements for temperatures above 1200°C. This comprises:

(i) two heating resistances cut in the same graphite cylinder (77 mm internal, 95 mm external diameter), 450 mm long, 150 mm for the supply terminals, and 300 mm for the resistance itself), and mounted in parallel;

Section of a cylindrical black body furnace

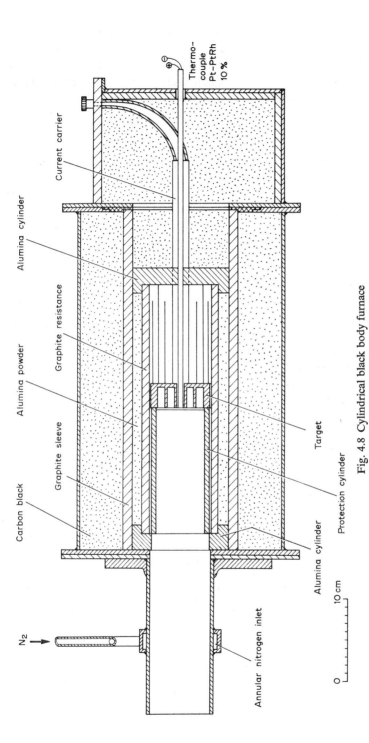

Fig. 4.8 Cylindrical black body furnace

(ii) a graphite target (73 mm in diameter, 30 mm thick) pierced with 31 holes 9 mm in diameter and 20 mm deep. This target forms the base of the furnace. It is located 150 mm from the entry;

(iii) a graphite cylinder 150 mm long and 73 mm in external diameter which covers the entry parts of the furnace to the target;

(iv) centring and insulation elements (between the target + inner cylinder and the resistance) made of alumina. The whole unit is centred on another graphite cylinder. Thermal insulation is obtained by means of carbon powder on the outside of the latter graphite cylinder, and alumina powder where the resistances are in contact;

(v) a sighting tube of 63 mm internal diameter which incorporates an annular inlet for nitrogen to prevent the graphite from burning. The flow of nitrogen to be injected is of the order of 40 l/min.

The emissivity of this furnace, taking into account the special form of the target, is about 0·985 for a graphite emissivity of 0·80. This furnace has the advantages of being able to withstand rapid changes of temperature (over 150°C per hour), of operating up to 1600°C, and of having a low consumption (3 kW at 1600°C) at high temperature. The main drawback is a short service life despite the injection of nitrogen, which is a considerable handicap.

The furnace is mainly suitable for calibrating the total radiation pyrometer.

Bibliography

1. MACADAMS, W. H. *Transmission de la chaleur*, Paris, Dunod, 1961.
2. BRAUD, Y. *Essais 0–15: Instruments de mesure utilisés pour l'étude des flammes incidentes vers la sole* (2 tômes). IJmuiden, I.F.R.F. Doc. nr. F 72/a/10, 1966.
3. CHEDAILLE, J. and MINEUR, J. M. *Essais 0–15: Transfert de chaleur à la sole dans le cas de flammes incidentes d'huile, avec air enrichi en oxygène* (2 tômes). IJmuiden, I.F.R.F. Doc. nr. F 31/a/41², 1967.
4. MAULARD, J. Fluxmètre thermique pour flux de rayonnement. *La Recherche Aéronautique*, **81**, pp. 37–38, March/April, 1961.
5. MINEUR, J. M. *Heat transfer measurements with the steel plug heat flow meter.* IJmuiden, I.F.R.F. Internal Report No. 76, 1965.
6. IVERNEL, MME A. Utilisation de l'oxygène pour le travail du verre—Mise au point d'un brûleur oxy-gaz. *Revue Générale de Thermique*, **7**, 78, pp. 637–645, June, 1968.
7. KOIZUMI, M. and ICHIKANA, M. Measurements of heat transfer to furnace walls using a thermo-electric heat flow meter. *Bulletin of the Japan Society of Mechanical Engineers*, **8**, 32, pp. 695–701, December, 1965.
8. GARDON, R. A transducer for the measurements of heat flow rate. *Transactions A.S.M.E.*, pp. 396–398, November, 1960.
9. ZINSMEISTER, G. E. and DIXON, J. R. Thermal flow meters and the effect of axial conduction. *Transactions A.S.M.E.*, pp. 64–68, February, 1966.

10. CABANNES, F. *Absorption et réflexions sélectives; leur importance dans certains problèmes de pyrométrie et d'échange d'énergie par rayonnement.* Silicates Industriels, pp. 501–508, 1964.

11. BEÉR, J. M. and CLAUS, J. The "traversing" method of radiation measurement in luminous flames. *Journal of the Institute of Fuel*, pp. 437–443, October, 1962.

12. Land Pyrometers Ltd., Sheffield, England. The Schmidt method of temperature measurement and the Land Twin Beam Radiation Pyrometer.

13. HEMSATH, K. H. *The precalculation of radiation emitted by pulverized fuel flames.* IJmuiden, I.F.R.F. Doc. nr. G 04/a/3², 1966.

14. GOUFFE, A. Considérations sur le rayonnement des flammes éclairantes. *Revue Générale de Thermique*, **19**, pp. 799–808, July, 1963; **20**, pp. 925–939, August, 1963.

15. KISSEL, R. R. *Additional equipment in use and instruments partly developed at IJmuiden.* IJmuiden, I.F.R.F. Doc. nr. Tb F 72/a/6 (supplement to Doc. nr. Tb F 72/a/4).

16. MICHAUD, M. *Facteur d'émission d'oxydes métallurgiques et réfractaires à haute température.* Thèse du Doctorat, Université de Paris, 1951.

Part 3

Measurements of Gas and Solids Concentration

Introduction

Measurement of the gas and solids concentrations is the most valuable of the various sources of information for the study and the control of industrial diffusion flames. It enables us:

(i) to study the mixture between the primary and secondary jets and the recirculation currents, therefore between fuel and combustion agent (possibly through the medium of a tracer);

(ii) to determine the process and kinetics of the combustion reactions;

(iii) to calculate the final output of the reactions from the unburnt fraction and the partial pressures of the gases which are present;

(iv) to control the smooth running of the furnace and possibly operate a system of regulation;

(v) to predict the radiation of the flames from the concentrations of solids, the granulometric distribution of the particles and the partial pressures of the radiating gases;

(vi) to control the atmospheric pollution due to fumes.

The measurement conditions vary quite considerably depending on whether pulverized coal, fuel oil or gas, is being burnt, and whether the sampling takes place in the burners, in the flame or in the flue, but one always comes up against the same major difficulties which are:

(i) the conservation of the initial characteristics of the sample;

(ii) the separation of the gaseous, solid and liquid phases;

(iii) the precise and if possible, rapid analysis of very complex gaseous mixtures;

(iv) the granulometric analysis of very fine particles from small quantities.

We shall study the successive stages of the measurement individually, from sampling and conditioning the samples to the analysis proper of the various constituents, at the same time stating the difficulties involved, and giving the solutions adopted at the IJmuiden station, finishing with a description of the equipment used.

Chapter 5

Sampling

When measuring concentration, the first operation is sampling, which is intended to extract a representative specimen of the mixture existing at the point of measurement and of preparing it so that it can be subjected to various analyses. The difficulty in this operation obviously depends on the characteristics of the sampled mixture—tendency to react (composition and temperature), number of phases (gas + solid or gas + solid + liquid at the beginning of fuel oil combustion), velocity and concentration of the particles, etc. Table 5.1 gives the possible limits of the principal characteristics of the environment present in the experimental furnaces at IJmuiden and encountered in the majority of industrial furnaces.

5.1 Sampling conditions

In order for the sample to be really representative, it is necessary that it should allow the inevitable large fluctuations in the diffusion flames to be integrated or possibly followed, and that apart from the partial pressures of the different gases present, it should conserve the concentration of particles and the granulometric distribution of these particles.

Method of sampling

In general, devices which operate by instantaneous capture of a certain volume of mixture, some droplets of oil or a few solid particles, are not suitable, since they furnish a sample which represents the state of the constituents at a given moment, whereas very large fluctuations which are a function of time are possible in diffusion flames (Fig. 5.1). Experience shows that the reproducibility of the mean measured values begins to be acceptable when the sampling duration exceeds one minute, and becomes good (to within $\pm 0 \cdot 2\%$ absolute, for the partial pressures of the gases when this duration reaches three minutes. These minimal sampling durations, which are valid for fuel oil and pulverized coal flames in the IJmuiden furnaces, may be quite different for other installations, particularly if cyclic phenomena take place in them.

In any case, whether the final object is to obtain mean values or continuous recordings, the measurement always calls for a sampling device with continuous suction, able to operate for at least several minutes.

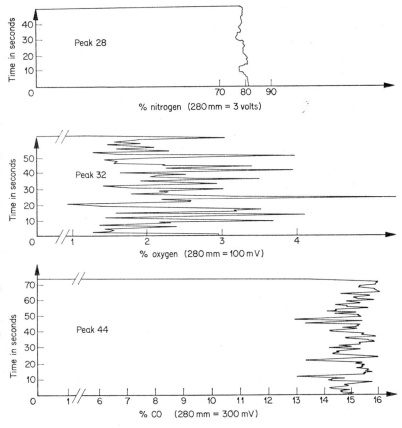

Fig. 5.1 Concentration fluctuations in a diffusion flame

Conservation of the chemical constitution of the mixture

The mixture which has been sampled is generally in the state of more or less rapid evolution (pyrolysis of pulverized coal particles, cracking of gas and fuel oil, various combustion reactions, dissociations and re-associations) which must be stopped quickly in order to conserve all the initial constituents with their original concentrations. This is achieved by a conditioning process of which the characteristics, that is, the final temperature reached and the rate of cooling must be fixed as a function of the kinetics of the reactions in progress, which in turn depends on the constituents present and their temperature. The stability conditions for the majority of the bodies being well known to the chemists, the selected final temperature is that at which the constituents no longer evolve, or at least very little.

On the other hand, a rigorous determination of the duration of the conditioning which involves the kinetics of phenomena which are in general very complex is not possible (1). At IJmuiden the gas samples are cooled to at least 300°C (2), a temperature at which the permanent gases, the pulverized coal particles and the soots effectively cease to evolve even in the presence of oxygen (the pyrolysis threshold for pulverized coal is about 500°C for a poor, and 350°C to 400°C for an enriched type (3)). We note, however, that reactions are still possible, either as a result of the presence of catalysts or because they involve very unstable bodies such as free radicles. On the other hand the cooling is often limited by the necessity to conserve vapours which condense easily.

In our case, since we do not seek to analyse heavy hydrocarbon vapours, the minimum admissible temperature is 150°C, a possible dew point for water vapour in the presence of traces of sulphur (mainly SO_3). If it is desired to maintain this temperature right along the suction line it is usually necessary to supply additional heating which can be done easily by electrical resistance.

As far as the period of cooling is concerned, it has been estimated empirically that it should be of the order of 3×10^{-3} seconds (2). By assuming that in the course of the conditioning there are no secondary reactions, and that the reactions taking place continue to do so only partially, the duration of the conditioning can be represented by a displacement of the point of sampling. The latter, at the average velocity of the gas (20 m/s) would be less than 6 cm in all cases; in other words the field of measured concentrations will be only slightly distorted in relation to that of the real concentrations since the complete combustion reaction in our furnaces takes place over a distance of several metres. Clearly this conditioning does not guarantee the detection of transient and very unstable states which may occur, mainly at the beginning of the flame. We have, therefore, to carry out in practice a conditioning at 300°C in less than 3×10^{-3} second, of the hot gases drawn continuously into the sampling line; the simplest method is to pass them from their entry into the probe, into a tube cooled by circulating water:

(i) just as for the other parts of the sampling circuits in contact with the inspired gases, one requires this tube to behave well at high temperatures, to be chemically inert and without catalytic effects on the oxidizing reactions; stainless steel meets these requirements satisfactorily; (ii) the conditioning tubes are made as shown in Fig. 5.2; an opening with

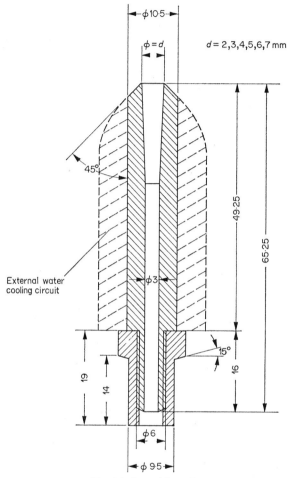

Fig. 5.2 Quenching tube

a diameter of 7, 6, 5, 4, 3, 2 mm, which varies according to the tubes, is joined to a 3 mm pipe. This arrangement makes it possible, whatever the velocity of the gases to choose a tube which samples isokinetically when drawing in a flow in the neighbourhood of 5 1/min. Thus the conditions of conditioning and filtration and the duration of the samplings always remain the same despite the wide range of velocities present in our furnaces;

(iii) a laboratory study (2) has shown that the cooling is sufficient if the length of the cooled tube exceeds 30 mm.

However, under these conditions, the gases remain much hotter along the axis of the device than along the wall, and the conditioning will only be completely effective when the tube is followed by a mixing chamber for the gaseous current (Fig. 5.3). This is realized by making the conditioning tube open into a filter in the form of a finger, preferably arranged perpendicularly to the current.

Thus gases at 1500°C, drawn in at 5 1/min are cooled to 200°C in less than 1/500 second (2).

Conservation of the concentration and granulometric distribution of particles

The particles suspended in a gas are subject to many forces: the force of gravity $P = mg$, the force of inertia $I = m\gamma$, the drag force $T = \rho V^2 C_x/2$ and finally the electrostatic forces.

> $m = \pi D_p^3 \rho_s/6 =$ mass of the particle of diameter D_p and of specific weight ρ_s
> $g =$ acceleration of gravity
> $\gamma =$ centrifugal acceleration of the particle $= V^2/R$
> $\rho =$ fluid density
> $V =$ relative velocity of the particle and the fluid
> $C_x =$ drag coefficient of the particle, depending on its shape

The solid and liquid particles behave differently, therefore, from gas molecules and in an undisturbed atmosphere, for example, fall at a velocity u which is given by

$$u = k_m g D_p^2 \frac{\rho_s - \rho}{18\mu} \quad \text{(Table 5.2)}$$

where k_m is a correction factor and μ is the dynamic viscosity of the fluid, assuming that Stoke's Law is applicable and that the particles are spherical. In our furnaces the particles are always small enough (diameter less than 100μ) and the gas velocities great enough (from 10 m/s to 80 m/s) for the drag force to be more significant than the weight: as a result, the trajectories of the particles in the jet where the stream lines are straight and where the deceleration of the gases is relatively slow, are practically identical with the stream lines, and their velocity remains very near that of the gas. These conclusions are not valid if for any reason the gases are abruptly decelerated or accelerated, or if the stream lines have a curvature which is not negligible; the inertia forces can then predominate over the other

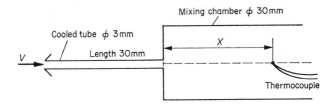

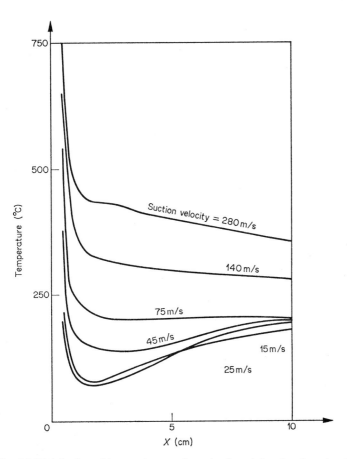

Fig. 5.3 Distribution of temperature on the axis of a mixing chamber placed after the quenching tube (gas temperature in the furnace 1550°C)

forces. This is precisely the case when a probe is introduced into the flow for continuous sampling of a flow of gas-particle mixture. Figure 5.4 illustrates the perturbations produced when suction is carried out by means of a thin-walled cylindrical tube, and makes it possible to formulate the following conclusions:

(i) the suction should be effected in the direction of the flow in order to limit distortion of the stream lines, which is very important in the case of suction in a static position, Fig. 5.4 (a) (**2**);
(ii) the rate of suction in a dynamic position should equal that of the gas flow (isokinetic), Fig. 5.4 (b). If the suction is too fast, the stream lines are deflected towards the axis of the suction tube, while the large particles, because of their inertia, continue along trajectories which are practically straight, and are not bunched. The error in the particle density will be on the low side and the proportion of fine particles too large. On the other hand if the suction is too slow, the stream lines are deflected towards the outside of the suction tube while the large particles which are not drawn in continue to enter. The density of particles will then be measured on the highside and the proportion of large particles is too large. Hence with a thin-walled tube it is necessary to carry out suction isokinetically in order to conserve both the particle density and the granulometric distribution.

When the rate of sampling V_m differs from the velocity V of the current, the measured concentration of particles C_m, defined as the ratio between the weight of solid and the volume of gas drawn in can be written, according to Badzioch (**4**) in the form:

$$(C_m - C)/C = \alpha(V - V_m)/V_m$$

or

$$C_m/C = \alpha \cdot V/V_m + (1 - \alpha) \tag{5.1}$$

where C is the actual concentration and α is a parameter depending on the particles (shape, size, density), on the fluid density, and the pattern of the stream lines in the neighbourhood of the probe. By multiplying equation 5.1 by the ratio SV_m/SV, where S is the suction section, we obtain

$$M_m/M = \alpha + (1 - \alpha)V_m/V \tag{5.2}$$

where M_m is the mass of particles collected per unit time and M is the actual mass flow of the particles in suspension.

If the particles are very fine $\alpha = 0$, and equation (5.1) gives $C_m = C$ whatever the rate of suction, which amounts to saying that the particles behave like fluid molecules. On the other hand if the particles are very large, $\alpha = 1$ and equation (5.2) gives $M_m = M$: the mass collected does not

— — — — Particle trajectories (diameter ϕ)

→ Stream line

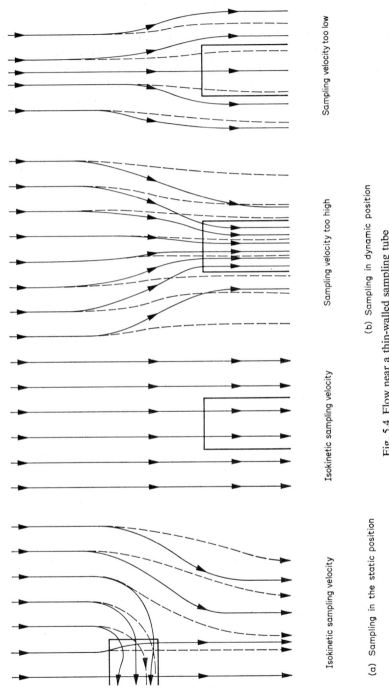

Isokinetic sampling velocity

Isokinetic sampling velocity

Sampling velocity too high

Sampling velocity too low

(a) Sampling in the static position

(b) Sampling in dynamic position

Fig. 5.4 Flow near a thin-walled sampling tube

depend on the rate of suction, and the exact concentration is then found by dividing this mass by the volume which one would have drawn in if the sampling had been made isokinetically. In the intermediate cases $0 < \alpha < 1$ the error will depend on the method used to calculate C_m.

The parameter α has been calculated theoretically for an ideal case by

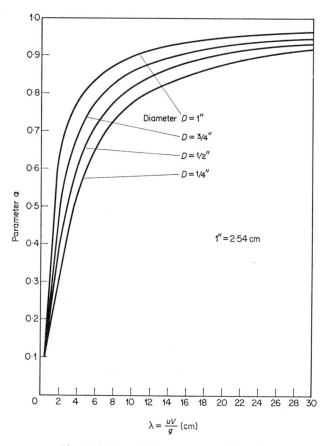

Fig. 5.5 Values of the parameter α (after (**4**))

Vitols (**5**), but it is more convenient to use the expression simplified by Badzioch, an expression which is valid subject to certain assumptions:

$$\alpha = 1 - e^{-L/\lambda}/L/\lambda$$

where $\lambda = uV/g$, with u being the velocity of free fall of the particles defined above when Stokes' Law is followed, and L is the length of the perturbation upstream of the suction tube which is a function of the tube geometry.

Figure 5.5 gives the value of λ calculated in this way.When the granulometry is extensive, the mean value of α is defined by $\alpha = \Sigma_i \alpha_i \Delta p_i$; where α_i is the value of α for particles of diameter i and Δp_i is the proportion by

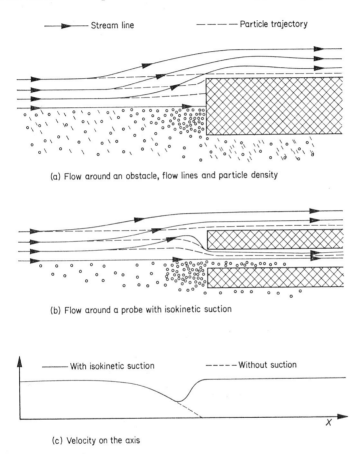

(a) Flow around an obstacle, flow lines and particle density

(b) Flow around a probe with isokinetic suction

(c) Velocity on the axis

Fig. 5.6 Flow near a thick-walled tube

weight of these particles. In our furnaces α, which is practically zero for the soot particles, reaches 0·60 for fly ash and exceeds 0·90 for pulverized coal at the burner outlet with a suction tube of 1 cm diameter. We maintain, therefore, that some exact corrections to the measured concentrations are possible when we know the velocity of flow, the rate of sampling and the granulometric distribution of the particles. Unfortunately these results are only valid for a thin-walled probe, while the probes which we are using have an appreciable thickness because they have to be water cooled.

In the neighbourhood of thick-walled probes the flow of fluid is intermediate between that described before for thin-walled probes, and flow round an obstacle (Fig. 5.6). Without suction the stream lines spread out ahead of the obstacle, and there is an accumulation of large particles in the same region which, because of the preponderance of their inertia, do not

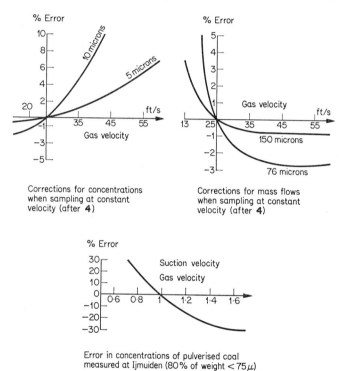

Corrections for concentrations when sampling at constant velocity (after **4**)

Corrections for mass flows when sampling at constant velocity (after **4**)

Error in concentrations of pulverised coal measured at Ijmuiden (80% of weight $< 75\mu$)

Fig. 5.7 Correction curves for non-isokinetic sampling

follow the curvature of the stream lines but strike against the stem of the obstacle, Fig. 5.6 (a). If a sample is taken isokinetically along the axis of this system, we shall obtain a tube of enlarged current having too high a concentration of large particles, Fig. 5.6 (b); the error in the concentration of solids and in the proportion of large particles will, therefore, be on the high side. The phenomena taking place have been illustrated for different shapes of probe heads by Walter (**6**) who also measured the velocities along the axis of the system, and obtained curves such as that shown in Fig. 5.6 (c).

The error incurred due to the thickness of the probe can be reduced in several ways:

(i) by suction at a higher rate than that of the flow; the ratio V_m/V to be used, which depends on the shape of the probe head, should be found by calibration;
(ii) by bevelling the leading edge of the probe in such a way that the large particles are always behind the opening in the suction tube and that they are thrown off after impact towards the outside of the probe. An experimental study carried out in England (7) on moderately thick probes has shown that an angle at the apex of the bevel between 15° and 20° will provide almost perfect efficiency;
(iii) by extending the cooled probe with a tube made of refractory material which samples just upstream of the strongly disturbed region.

Thus, when it is necessary to collect relatively large particles, the majority of the sampling devices must be calibrated before use, particularly if strictly isokinetic suction is not possible. In Fig. 5.7 we give some examples of correction curves as a function of V_m/V, and of the granulometry of the particles, and the method of calculation employed.

Filtration of particles

The calculation of the concentration of solids requires the sampled solids to be weighed and the total volume drawn in to be measured; that is, the complete separation of the two phases by filtration. The weight of sampled solids being quite small as a rule, it is essential to avoid losing any as a result of deposition ahead of the filter by placing the latter in the head of the probe, immediately following the conditioning tube, an arrangement which is, moreover, favourable for the conditioning of the gases, as we saw on page 96. The choice of the filtration method obviously depends upon the measurement conditions: the granulometry of the particles, the concentration of solids, the rate of suction, the available space, the characteristics of the gas (temperature, humidity, density), etc.

We find that we need a filtering system which can withstand high temperatures (300°C) and humidity, is easy to install, of small volume and possesses good mechanical properties in order to resist strong denting and be able to stand numerous handlings without damage. These requirements rule out paper filters which have, however, the advantage of retaining very fine particles, as well as electrostatic filters which are difficult to miniaturize. We have adopted filters made of calcined bronze with a maximum pore diameter of 10 or 5 microns (they can also be made with passages of the order of a micron but these cause a considerable loss of charge). The reduction of their diametral size is obtained without loss of effective filtering surface which must be of the order of 20 cm², by making them in the form of a hollow finger capable of collecting 3·5 g of pulverized coal. The rate of passage of the gases does not exceed 15 cm/s for the maximum possible flow of 8 Nl/min.

A series of tests undertaken in fuel-oil flames has made it possible to

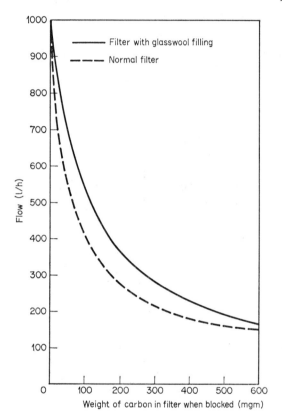

Fig. 5.8 Blockage of sinter bronze filters (5μ to 10μ pores)

determine the effectiveness of this type of filter for collecting the soots ($\phi < 1\mu$):

(i) The weight of solids traversing the filter is very small whatever the duration of sampling; with the filter pore diameter of 5μ, this weight is not measurable, and with 10μ passages it amounts at the most to a few milligrammes. (With longer filters the loss of soots becomes too large.) This surprising result is explained by the fact that the pores of the filter clog up very quickly after a few seconds of suction. Moreover, filters in use, already partially clogged, give less loss of soots than new filters.

(ii) The complete blocking of the filter (infinite loss of charge) occurs (Fig. 5.8) with less solids the more the rate of sampling is raised. On the one hand this result confirms the existence of a filtering film formed by the sampled particles, which is the more compact the greater the rate of sampling, and on the other hand it can explain the very slight increase in the measured concentrations as a function of the rate of sampling found in the course of the tests (8). Plugging the filter with glass wool has the effect of delaying the clogging since it retains the particles to some extent; it enables more solids to be recovered.

(iii) Considering the fluctuations of the flame, the reproducibility of the measurements is excellent (better than $\pm 5\%$) within the normal limits of variation of the rate of suction, the entry diameter of the conditioning tube and the concentration of soot.

The results for pulverized coal flames are even more favourable because on the one hand the losses are negligible because of the larger dimensions of the particles, and on the other there is no clogging, so that it is possible to collect sufficient particles for a final analysis.

In order to calculate the concentration of solids the filter is weighed before and after each sampling on a Mettler electrical balance which has an accuracy of ± 0.5 mg (a scale of 0–200 g).

In order to avoid losses and to facilitate operations the combined conditioning tube-filter unit is weighed dry and cold. Drying is carried out by passage for half an hour through a vacuum stove heated to 105°C, and cooling is effected in a desiccator at ambient temperature.

5.2 Probes and sampling circuits

In this section we give a description of the sampling equipment used at IJmuiden which comprises:

(i) a cooled probe which ensures the conditioning of the mixture drawn in and the protection of the suction line as far as the doors of the furnace;

(ii) a suction line in which the filtration, conveyance and conditioning of the sampled mixture is carried out;

(iii) a pump and devices for measurement, monitoring and control of the suction rate;

(iv) a subsidiary circuit either directly to the analyser if this is of the continuously operating type, or to a gasholder for keeping a sample of the gas.

Sampling probe

There are different versions according to whether the sampling takes place in the flame itself (standard probe), in the coal before ignition, or in the oil jet before vaporization (special probes).

Standard probe. The first probe that was made is shown in Fig. 5.9 (a): a water jacket of special shape due to the necessity of sampling in a dynamic position ensures on the one hand the conditioning of the mixture under the anticipated conditions (9), and on the other, the protection of the filtering and suction equipment. The effectiveness of the cooling of the conditioning tube T is ensured by the judicious arrangement of a block of metal which forces the water, coming from a special supply line, to circulate at great speed along the exchange surface. The probe, and the sampling circuit inside the latter, are made of stainless steel. The conditioning tube T, 50 mm long overall, and with an external diameter of 10 mm, protrudes about 6 mm from the water jacket and its stem is bevelled for the reasons given in paragraph (ii) on page 105.

The connection of T to C is made by an annealed copper joint with a conical seating, while the support A locks the brass ring (in which the filter F is fitted) against a seating arranged in C by means of a silicon rubber O-ring joint, which retains its properties up to 180°C and ensures gas-tightness. Before each measurement this arrangement is tested systematically under a vacuum of −640 mm Hg: after the opening in T is closed no leakage should be found. Temperature measurements with exposed thermocouples carried out at three points in the filter gave the results shown in Fig. 5.10: since the gas temperatures are certainly too low it appears that the conditioning was slower than envisaged in the first place despite the filter being placed perpendicularly to the conditioning tube. This probe was modified some years ago, as shown in Fig. 5.9 (b), in order to

(i) have available a detachable head mounted in the same cooled probe support as the standard Prandtl tube;
(ii) sample in the probe extension, that is, without axial time-lag in relation to the points of measuring speeds and temperatures;
(iii) reduce the handling time by facilitating the mounting and recovery of the conditioning tube-filter assembly;
(iv) reduce the perturbations at the point of sampling by cutting down the diameter and improving the aerodynamics of the probe head.

The filter is fitted in the extension of the conditioning tube and gas-tightness is provided by two O-ring joints J_1 and J_2 in silicone rubber, locked by the locking screw E on the device. Following the filter the suction tube

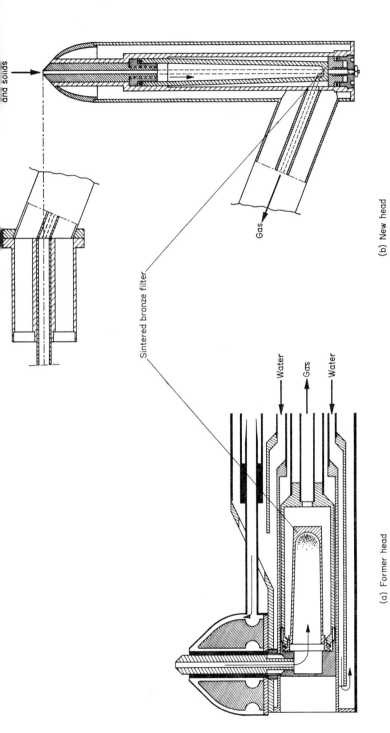

and solids

Sintered bronze filter

Water

Gas

Water

(a) Former head

Gas

(b) New head

Fig. 5.9 Gas and solids probe

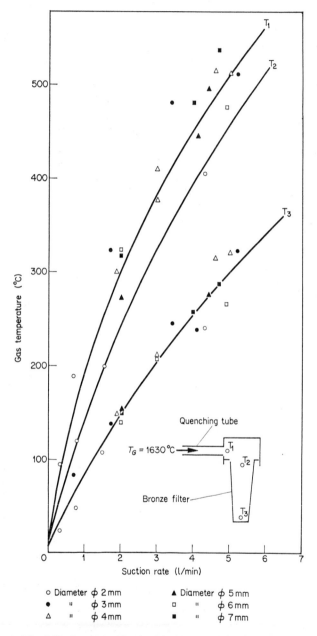

Fig. 5.10 Gas quenching in the old type of sampling probe

is cooled directly by the current of water in the head of the probe, so that condensation of the water vapour may take place. If necessary this inconvenience can be avoided by fitting an intermediate jacket. The conditioning of the gas was studied on this occasion (11) by introducing a miniature suction pyrometer in the filter housing (Fig. 5.11) so that the temperatures are measured with an accuracy of a few degrees. The results obtained (Fig. 5.11) show that the conditioning is just as fast as in the old probe. The most favourable suction flow which makes possible both a reasonably short sampling time and a sufficiently fast conditioning is in the region of 4 Nl/min.

Special probes. The filters in the preceding probes are too small to allow of sufficient sampling time in certain coal jets where the particle density is very high. On the other hand the dimensions of the probe heads are too large and cause the jets to break up with a resulting modification of the equilibrium of the furnaces. A special probe (Fig. 5.12) has been designed, therefore, for sampling in coal jets (12) which has a cooled suction pipe of 12 mm external diameter opening into a large capacity filter (35 g of pulverized coal, say ten times the volume of the standard probe filter), set off laterally by 200 mm which makes it possible to investigate coal jets without these penetrating the large probe. The suction tube has an opening of 2 mm and an interior diameter of 4 mm. A probe of this kind can operate effectively while the proportion of water vapour remains small, otherwise there is blocking. The conditioning has not been monitored since this probe is intended to work ahead of the flame where there is still no combustion, but the previous results enable us to confirm that the gas is cooled too fast.

Another probe, at present under development, is designed to take samples in a liquid fuel jet. The conditioning tube opens tangentially into a cyclone on the walls of which the fuel droplets are deposited by centrifugal force. A filter is placed at the outlet of the cyclone in order to collect the soot particles.

Sampling circuits

These circuits are the lines and the auxiliaries which ensure the transport of the sampled gas from the exit of the cooled probe to the analysers or to the storage container. The design of these circuits obviously depends on the performance required and the object aimed at.

Standard circuit. This ensures the functioning of the sampling probes described above, that is, the monitoring and precise control of the suction

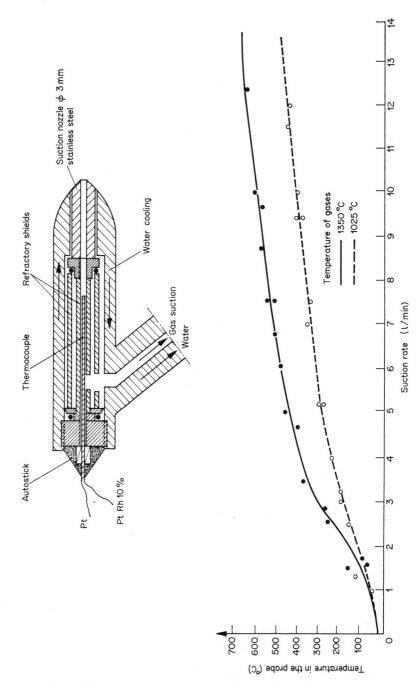

(a) Temperature at the exit of the suction nozzle

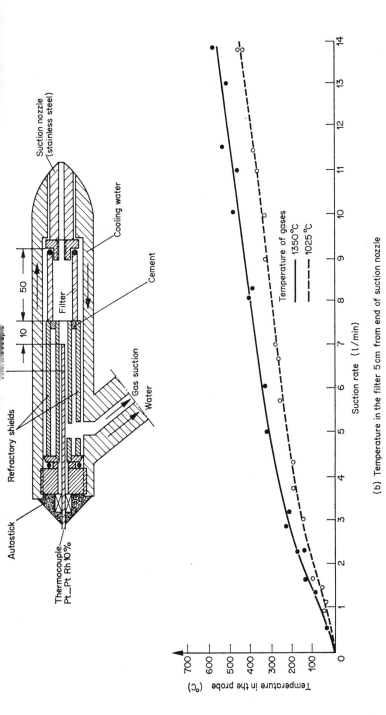

(b) Temperature in the filter 5 cm from end of suction nozzle

Fig. 5.11 Conditioning of the gas in the new sampling probe

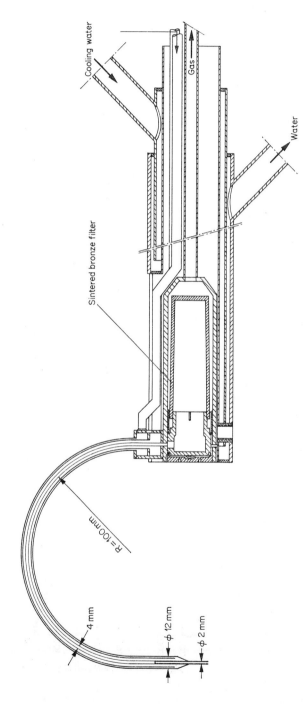

Fig. 5.12 Miniature gas sampling probe

low, the measurement of the total volume drawn in and the storage of a representative specimen of the average gas mixture.

It comprises, from the outlet of the probe (Fig. 5.13), of

i) a flexible coupling of thick rubber to control the displacement of the sampling probe along the furnaces;

ii) a circulating water condenser (the water which is collected cannot be weighed with sufficient accuracy to give the water vapour content in the gas drawn in);

iii) a pilot filter in a transparent tube in order to check the quality of the filtering in the head of the probe;

iv) a rigid connection up to the control panel;

v) a control panel as shown in Fig. 5.14, intended for two probes working simultaneously: one samples inside the furnace while the other is serviced outside (changing the filter). The gate V_4 serves to change the probes over to the two systems employed for:

(*a*) checking the air-tightness of the probe. The conditioning tube being closed off, a vacuum of 640 mm Hg is produced in the suction circuit from probe 1 by means of pump 2. The gate V_7 is then closed and the column of mercury in the manometer 2 should not fall; and for

(*b*) sampling the gases and solids: the pump 1 is brought into action at the moment when the head of the probe, introduced perpendicularly to the streams of fluid in order to avoid the re-entry of large particles, is set in the direction of flow. The suction flow governed by the gate V_1 is monitored by the rotameter 1, the gate V_2 being half-open and the gate V_3 closed. The pressure drop between the head of the probe and the panel is given by the manometer 2. After the suction time required to purge the dead volume of the probe (1·5 to 2 litres of gas), the gate V_3 is opened in such a way as to allow the gases to enter the empty jacket E made of gas-tight flexible plastic, enclosed in a box B, itself gas-tight and supplied with a drain hole through which the air displaced in the course of inflating E is returned to the main circuit in order to be taken into account by the volume meter. The flow and the perssure of refilling E are adjusted by the gates V_2 and V_3 and measured by the rotameter 2 in such a way that the filling time is several minutes in order to ensure that the gas sample stored in this way is representative (see the first section in 5.1). Finally, the water volume meter shows the total volume sampled which is essential to know in order to calculate the concentration of solids.

We use diaphragm pumps with which it is possible to draw in up to 10 l/min. A former study showed that samples containing hydrogen can be

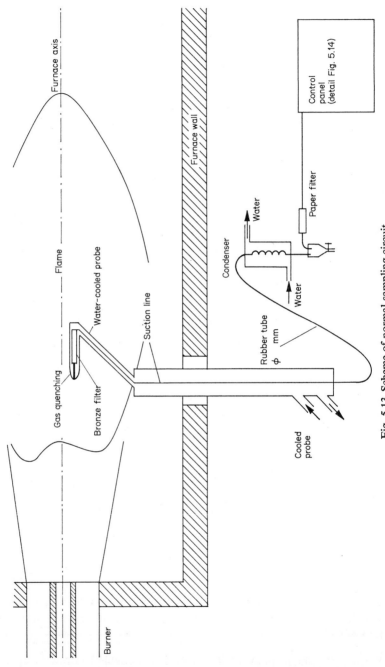

Fig. 5.13 Scheme of normal sampling circuit

Front view of table

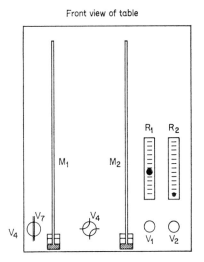

System to test
for probe 1   Sampling system

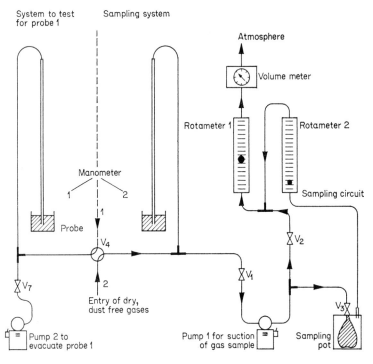

Fig. 5.14 Flow sheet for gas sampling

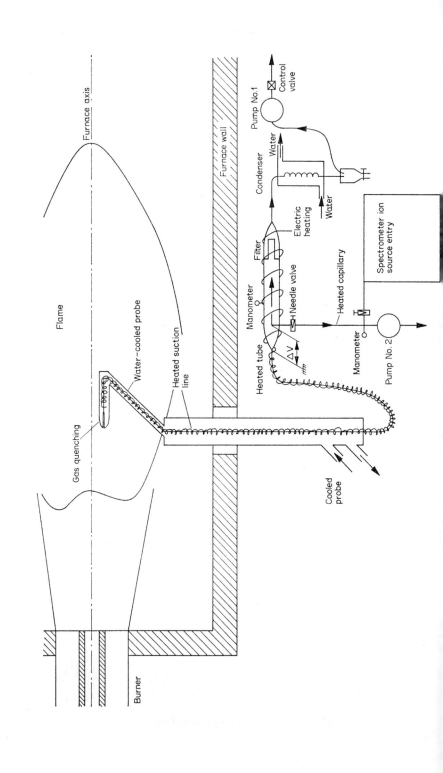

Flame

Furnace axis

Gas quenching

Water-cooled probe

Heated suction line

Furnace wall

Burner

Cooled probe

Heated tube

Spectrometer ion source entry

Heated capillary

Manometer

Pump No. 2

ΔV

Needle valve

Manometer

Filter

Electric heating

Condenser

Water

Water

Pump No.1

Control valve

kept for 24 hours in our gasholders without an Orsat analysis showing any appreciable loss by diffusion through the plastic membrane. However, these losses exist because they are detected by chromatography, and that is why it is strongly advised that the samples should be analysed inside an hour following the sampling or at the latest a few hours afterwards.

With these circuits we can replace the sampling flask by a continuous analyser (oxygen paramagnetic analyser, infra-red analyser, spectrometer). It is essential therefore to take care to eliminate completely any particles or water vapour by passing the gases through a drying tube (containing P_2O_5 or $(ClO_3)_2Mg$) followed by a filter. The response time of the line for 99% of the signal is then better than 15 seconds. If this delay is too long for the particular study in hand it is necessary to use circuits which are modified in the way described below.

Fast response circuit. This is shown in Fig. 5.15, and comprises a sampling probe of which the conditioning tube is followed immediately by a pipe heated to 150°C leading directly into the branch to the analyser, which in this case is a mass spectrometer (**13**). The main suction line is then fitted, like the one above, with a filter, a condenser, a pump, and a regulating gate for the flow or the pressure measured on the right of the branch to the analyser. The diameter of the heated pipe must be small while permitting turbulent flow. A diameter of 2 mm has been used for measurements in fuel-oil flames and one of 3 mm to 4 mm for coal flames because of possible high particle concentrations in the first part of the furnace. The main difficulty is in preventing the particles from entering the tube leading to the analyser.

It would seem possible to achieve this with an arrangement similar to that shown in Fig. 5.15, in which the particles are repelled by electrostatic force. With a distance of 10 metres between the sampling point and the analyser we obtain 90% of the input signal at the output of this suction line in a time of 0·5 seconds.

Chapter 6

Gas Analysis

We are mainly concerned with the analysis of gas mixtures which are sampled from different regions of the flames of pulverized coal, fuel-oil or gas, mixtures which contain many constituents in the most complicated cases (nitrogen, argon, carbon dioxide, oxygen, water vapour, carbon monoxide, hydrogen, methane, hydrocarbons and various traces), and whose concentrations vary in an unforeseeable manner over wide limits as one moves from one sampling point to another (Table 6.1). Such conditions make any precise and quick measurement difficult, despite the fact that nowadays we have available a great variety of analysers using the most varied properties of the gases for their indentification and measurement of concentration.

In this chapter we shall describe the principal devices which we have tested or are using at IJmuiden, these being numerous and varied enough to meet the majority of industrial requirements; manual or automatic analysis, continuous or discontinuous, complete or partial, absolute or merely reproducible measurements, trace detection, etc. For readers who are not familiar with these problems we felt it would be useful to recall the fundamentals of the theory needed to understand the principles and the mode of operation of the instrument.

6.1 Orsat analyser

This is the simplest and mose widely used analyser. It is based on chemical principles and operates manually by measuring the volume of the sample to be analysed in a graduated burette, before and after a passage through an absorbing reactant which is selected for one constituent only, the operation being repeated for each analysable constituent. There are numerous versions of this instrument which differ in the reactants employed, the design of the apparatus or the direction for using it.

The one we use—the "Complete Orsat"—is intended for the analysis of the following six gases: O_2, C_nH_m (unsaturated hydrocarbon) CO, O_2, H_2 and CH_4, in less than 20 minutes with an absolute accuracy which is comparable with the reproducibility of the samples taken from the flame (0·2%).

Choice of the reactant

The method of preparation and the properties of a large number of reactants are given in the literature (14) from which we have selected those which are simple and convenient in use and which give the fastest absorption (15):

(i) absorption of carbon dioxide; a solution of 30% by weight of potash (KOH) or a solution of 20% by weight of caustic soda (NaOH). These reactants take up equally SO_2, SH_2 and C_6H_6, gases which must be eliminated before the analysis, in order to avoid estimating them as CO_2. It is worth remembering that 1 cm³ of solution absorbs about 50 cm³ of CO_2 and that the reaction is very fast but never complete; the residual may amount to 0·1% (14);

(ii) absorption of unsaturated hydrocarbon by fuming sulphuric acid with 0·6% by weight of silver sulphate (or vanadic acid) accelerator. This reactant has two drawbacks; it liberates a little SO_2, a gas which should be absorbed by passing through a potash solution (this means that the CO_2 analysis precedes that of the C_nH_m), and it is self-diluting by taking up water vapour, so that it has to be made up from time to time by the addition of fuming sulphuric acid. Other reactants, such as bromide water or a solution of $CaCl_2$ + $CaBr_2$ + Br, are much more awkward to use;

(iii) absorption of oxygen by pyrogallol, consisting of 9 g of pyrogallic acid dissolved in 36 cm³ of distilled water and mixed with 225 cm³ of a 60% solution of potash. This reactant oxidizes in air (the surface of the liquid exposed to the air is protected by a film of oil which is immiscible with the solution) and in light (the solution is made up shortly before it is to be used); it must be changed as soon as brown coloured grains appear on the walls of the flask. One cubic centimetre absorbs about 12 cm³ of oxygen, but also takes up a little CO which it gives up after a certain time. In addition, white phosphorus in water, or solutions of hydrosulphite or chromium chloride are often used;

(iv) oxidation of carbon monoxide to carbon dioxide by acid iodine pentoxide prepared as follows: 120 g of powdered I_2O_3, 300 cm³ of 20% sulphuric acid, and 150 cm³ of sulphuric acid, are kept 24 hours in a porcelain grinder charged with glass balls ($\phi \simeq 10$ mm and $\phi \simeq 3$ mm). This solution is then washed in a beaker with 150 cm³ of fuming sulphuric acid (if it is too thick after this a little more sulphuric acid can be added). Obviously this reactant absorbs the C_nH_m and produces a little SO_2. The gases are passed through the potash again in order to remove the CO_2 which is produced, which makes it necessary to analyse the CO after the CO_2 and the C_nH_m. Copper chloride in acid, neutral or ammoniacal solution is also frequently employed for direct absorption of the CO;

(v) combustion of hydrogen and methane. There is no known reactant for methane, while hydrogen is absorbed by palladium, but only after a very long time (10 to 30 minutes according to (**14**)). Provided that the mixture does not contain any other combustible, these gases can be analysed by proceeding as follows: the gases are burnt by passing through a platinum resistance eudiometer heated electrically (or in a platinum or silica tube raised to 800°C), in the presence of excess air, and the contraction of the total volume ΔV is measured, as well as the volume V_{CO_2} of carbon dioxide and by-products by passing through potash. The volume of methane in the original mixture is equal to V_{CO_2} and that of hydrogen is given by

$$V_{H_2} = \tfrac{2}{3}(\Delta V - 2V_{CO_2})$$

(the water is recovered in the liquid form).

In order that the oxygen should be excess it is usually necessary to retain only a small proportion of the volume remaining after the absorption of the four major gases so as to have sufficient combustion air. The result is that the errors incurred for methane and hydrogen are multiplied by the inverse of this proportion.

Description of the equipment

The instrument we use (Fig. 6.1 (a)) is fitted inside a wooden box which can be closed by sliding panels during transport and storage. The equipment consists of a 100 cm³ glass burette, graduated in 1/10 cm³ over the first 25 cm³, immersed in a tube filled with water intended to provide insulation against external temperature fluctuations during the measurement and joined at the base to a detachable container R by means of a flexible rubber tube. This container, R, which is filled with water coloured red by methyl orange and contains 5% sulphuric acid in order to limit absorption (water with salt or saturated with $CaCl_2$ and acidified with HCl will do as well), serves in accordance with the law of communicating vessels to expel or admit the gas into the burette by varying its level, and to restore the gas to atmospheric pressure by aligning the two liquid levels. The upper part of the burette is connected to ambient air and to Lomchakov type

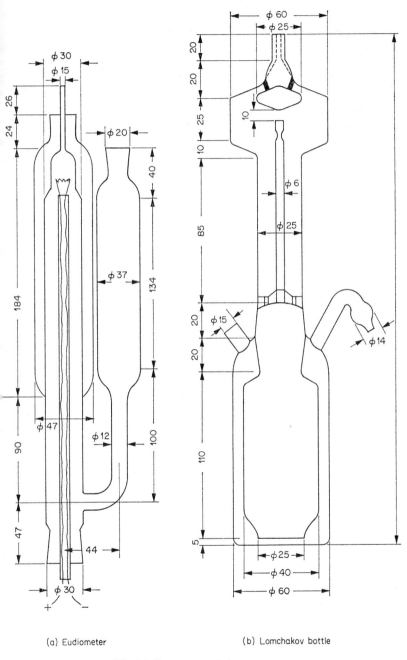

(a) Eudiometer   (b) Lomchakov bottle

Fig. 6.1 Orsat gas analysis apparatus

absorbers by means of capillary tubes of 1·5 mm diameter, of negligible dead volume and closed by three-way cocks. These absorbers, Fig. 6.1 (b), which contain 250 cm³ of reactant of which a small part is necessary for the absorption reactions operate by dispersion of the reactant into the gas being analysed in the form of a cloud of fine droplets, a procedure which by creating a large reaction surface makes it possible to reduce the duration of the absorption.

Finally, a platinum resistance eudiometer, Fig. 6.1 (c), heated electrically, serves to burn the hydrogen and methane in the presence of excess air; sealing liquid consists of water containing $10\%H_2SO_4$.

Method of operation

Since the reactants are not ideally specific, the analysis is carried out in the following order: CO_2, C_nH_m, O_2, CO and finally $H_2 + CH_4$. Two absorbers are added in order to provide nitrogen from the air for filling the dead spaces (the reactants must never reach the cocks). After this is done, 100 cm³ of gas for analysis is admitted to the graduated burette. This volume is then expelled into the upper part of the first absorber (containing potash solution) by the action of the reservoirs. A quick compression of the rubber bulb connected to the olive-shaped tube, Fig 6.1 (b), forces the reactant into the 6 mm central tube, projects it against the conical member and disperses it in the form of a cloud of fine droplets into the gas being analysed. The constituent is therefore rapidly absorbed (15 s for CO_2) and when the reaction is completed the gas is returned to the burette for the remaining volume to be measured. These operations are then repeated for the other constituents. The time for a complete analysis is about 20 minutes.

Sources of error

(*a*) Leakages: it is essential to ensure that the cocks are completely gas-tight by suitably lubricating them with a silicone grease (in very small amounts, never to excess).

(*b*) Incomplete absorptions; these originate either from the use of poor quality reactants or from carrying out the analysis too quickly; the duration and the method of operation should be fixed accurately by preliminary tests.

(*c*) Measurements of the gas volumes. We have first of all the reading error, estimated at 0·05 cm³ when the burette is graduated in 0·1 cm³, and then there are errors due to drops of liquid being retained on the wall possible variations of temperature, poor adjustment of the liquid levels etc. It should be noted that we are measuring a gas saturated with water vapour which does not constitute a source of error if the temperature remains constant since the original reference volume is also saturated. The total error in measuring each volume can be taken as 0·1 cm³, which means that with 100 cm³ being introduced, the proportion by volume of the absorbed constituents under the best conditions is known to within 0·2%. As we have seen above, the errors in the cases of CH_4 and still more of hydrogen are higher. Finally, the maximum error for these six constituents is that associated with the measurement of the proportion of inert gases (in our case nitrogen + argon) obtained by taking the differences. This is obviously greater than 1%.

Conclusion

The Orsat analyser which we use is easy to handle and enables the four major gases (CO_2, C_nH_m, O_2, CO) to be analysed in about 10 minutes and a complete analysis in less than 20 minutes. Its accuracy which is acceptable for CO_2, C_nH_m, O_2 and CO ($\pm 0·2\%$ absolute), taking into account the reproducibility of the samples taken in our flames, is not good enough for H_2, CH_4 and the inert gases. An improvement in this accuracy does not seem possible without considerable lengthening of the analysis time.

6.2 Chromatography

This is a semi-automatic or automatic analyser based on physico-chemical principles (Fig. 6.2). A small volume of the gas to be analysed is introduced by means of a carrier gas into a column filled with a stationary adsorbing phase which reconstitutes the different constituents separately; these are then analysed quantitatively (and if necessary, qualitatively) by suitable instruments. A steady improvement in the quality of the stationary phases filling the columns and of the technology of the auxiliaries (device for introducing the gas, flow controller, detectors, etc.) has enabled chromatography to extend its range of application continuously and has already made it the most complete, and most widely used analyser. It is probable, moreover, that in the next few years it will replace the majority of the discontinuous analysers based on chemical principles like the Orsat, in view of its ability to separate the most diverse constituents of very complex mixtures under perfectly reproducible conditions. In this Section we shall only deal with the possibilities it offers for the analysis of combustion gases containing H_2, O_2, N_2, CH_4, CO, CO_2 and the light hydrocarbons up to C_4).

However, as this method of analysis is relatively new and many engineers

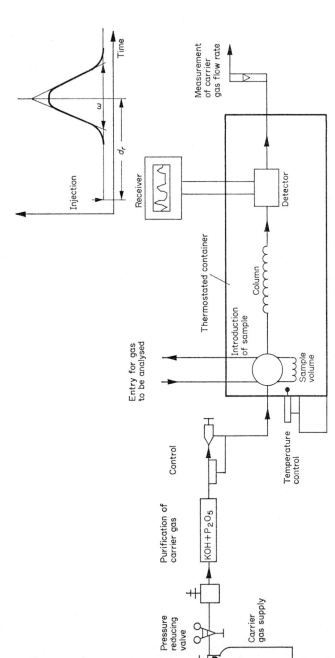

Fig. 6.2 Principle of a chromatograph circuit

ıre not well acquainted with it, we thought it advisable to recall the essen-
ial features and the possible variants before describing the instrument
ʋhich we have installed.

Displacement of a constituent in a chromatographic-column

The stationary phase is either a solid adsorbent powder or a powdery inert
ɔlid base impregnated with a solvent liquid. When a small volume of a
ɡaseous constituent is abruptly injected into a current of carrier gas and
ɟrawn by it into the column, it establishes an equilibrium between the
raction of constituent adsorbed and that remaining in the gas phase. Thus
he constituent (or solute) held back is displaced in the column at a mean
peed (displacement of the concentration maximum) which is a function
ɔn the one hand of the velocity of the carrier gas, and on the other of its
ɔroperties in relation to the stationary phase. It then reaches the end of
he column after a certain time called the time of retention, characteristic
ɔf the solute and of the different control variables, in the form of a peak
ɔf concentration of gaussian form, which one tries to obtain as narrow as
ɔossible.

The process being similar to that of a distillation column, we call the
hickness of the column w required for the moving phase coming from it
ɔ be in equilibrium with the stationary phase contained in it the height
ıquivalent to a theoretical plateau (Hept); it has been shown **(16)** that
he number of theoretical plateaux r in a column can be calculated from
ı chromatogram, provided that certain assumptions are justified—negli-
ɡible dead space, instantaneous injection of the solute, etc.

$$r = (16 d_R/w)^2 = L/h$$

ʋhere L is the length of the column, d_R is the product of the retention time
ınd the rate of unrolling the paper in the recorder, and w is the distance
ɔetween the points of intersection of the base line and the tangents to the
ɔoints of inflection of the peak (Fig. 6.2). As a result:
a) for equal retention times, the columns having the greatest number of
ɔlateaux give the narrowest peaks, and therefore the best separation;
b) the width of the peaks is proportional to the time of retention; in a
hromatogram the first peaks are the narrowest;
c) h is independent of the mass of solute introduced so long as there is
ɪo excess charging (saturation) of the column;
d) lengthening of the column increases the number of plateaux, and there-
ʋore improves the efficiency at the expense of the retention time.

According to van Deemter's classical formula, which gives the theoretical expression for h, the latter depends on the linear velocity of the carrier gas, *i.e.* on its pressure:

$$h = 2\lambda d_p + 2\sigma D_m/U_m + \frac{8}{\pi^2} \frac{K'}{(9 + K')^2} \frac{e_f^2}{D_f} U_m$$

where:

λ = dimensionless parameter which takes into account the irregularities in filling the column;

σ = correction factor (<1) which takes into account the divergencies in the path followed by the gas in the column;

d_p = diameter of the solid particles in the fixed phase;

D_m = coefficient of molecular diffusion in the moving phase;

U_m = linear speed of the carrier gas;

$K' = \dfrac{KF_f}{F_m}$ = partition coefficient $\times \dfrac{\text{fraction of the fixed phase volume}}{\text{fraction of the moving phase volume}}$

e_f = mean thickness of the fixed phase film;

D_f = coefficient of molecular diffusion in the fixed phase.

Thus h varies from the beginning to the end of the column with the pressure of the carrier gas and, to the extent that the lengthening of the column causes an increase in the pressure upstream, the number of theoretical plateaux does not depend linearly on the length of the column, other things being equal;

(*e*) the diameter of the column has no effect on the separation but it does determine the maximum charge, that is, the volume of gas to be analysed which can be introduced. It should be noted, however, that diameters which are too large cause distortion in the concentration front, and therefore a loss of efficiency. We shall see later that the volume of the sample introduced into the column is fixed as a function of the sensitivity of the detector;

(*f*) van Deemter's formula can be written as

$$h = A + B/U_m + CU_m$$

where A, B and C are constants for a given column. The function $h(U_m)$ shown in Fig. 6.3, has a minimum; that is, there exists one value, and one only, of the flow of carrier gas for which the efficiency of the column is maximum. Since we can measure h experimentally it is easy to obtain the optimum value of U_m; it is found in general that the curve of $h(U_m)$ is flat in the neighbourhood of the minimum, so that it is possible, without loss of efficiency, to choose a higher flow of carrier gas in order to reduce the

analysis time, the time of retention, of course, being inversely proportional
to this flow;

(g) Wet and Pretorius (**25**) have shown, subject to certain assumptions,
that

$$h = A' + B'/T + C'T$$

that is, there is also a temperature which is optimum for the efficiency
of the column; a temperature which, like U_m, can be determined experimentally. On the other hand the logarithm of the retention volume varies

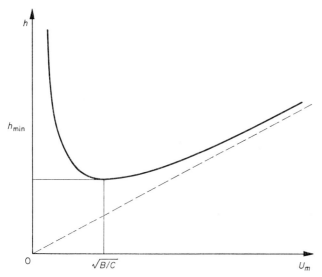

Fig. 6.3 Height h of a theoretical plateau as a function of the flow of
carrier gas U_m

linearly as an inverse function of the temperature (Fig. 6.4). (The retention
volume is the volume of carrier gas which flows during the passage of the
solute through the column.) Therefore in order to reduce the time of
retention it is necessary to operate at a temperature which is as high as the
curve $h(T)$ will allow.

Finally, we are well aware of the effect of the different parameters (flow
of carrier gas, temperature, length and diameter of the column) on the
efficiency of the column for a given solute and on the time of retention
of this solute, but our main interest is, of course, the analysis of a mixture
of several gases, which necessitates the construction of a column which
completely separates the peaks of each of the constituents in as short a
time as possible. The constituent is then identified by the time of retention

and the concentration measured by a binary detector with a fast response
which is located at the outlet of the column.

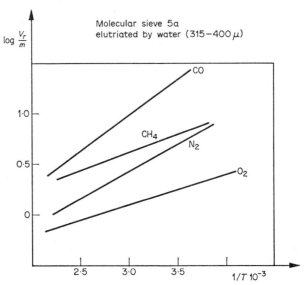

Fig. 6.4 Influence of temperature on the retention volume (after (**21**))

Choice of the column filling. In our case it is a question of separating a
mixture of the so-called permanent gases H_2, O_2, N_2, CH_4, CO, CO_2 and
light hydrocarbons (from C_2 to C_4). This separation can be achieved with
a solid adsorbing phase:

(i) a molecular filter made of silicates having pore diameters comparable
with that of the molecules, separates the permanent gases H_2, O_2, CH_4
and CO quite well, but absorbs CO_2 and hydrocarbons very strongly (a
low temperature this is an irreversible process);

(ii) active charcoal (coconut charcoal) behaves similarly to molecular
filters;

(iii) silica gel is much used for separating CO_2 and the light hydrocarbons
but does not separate the permanent gases satisfactorily at ambient
temperature (**17**);

(iv) alumina separates hydrocarbons quite well but has the property of
adsorbing CO_2 irreversibly (**17**).

The quality of these solid phases depends mainly on the method of preparation and activation.

(i) the specific surface of the powder and the fineness of the pores play a
very important part. Short, fine-grain columns separate better than

those which are long and coarse-grained, and the efficiency is greater when the granulometry is restricted.

(ii) no less important is the method of activation which enables the stationary phase to be freed of strongly adsorbed gases (particularly water vapour) by heating to a high temperature in a dry atmosphere or under vacuum.

(iii) the activity of the column falls during use if it is supplied with gases which are absorbed irreversibly under the particular operating conditions.

It is necessary, therefore, to see that the chromatograph is supplied with purified carrier gases, and if need be, protect the column by a pre-column which absorbs the harmful constituents irreversibly. In all cases only perfectly dry gases should be used. When this is possible, the life of the column is prolonged because of the reduction in volume of the samples which are introduced.

We realize that the separation of several solutes will take place easily if they are adsorbed differently by the stationary phase at the operating temperature, but that the duration of the analysis will be all the longer the more marked the differences in volatility. This is exactly the position we face; the range of boiling points of the various constituents which are to be separated is very wide. It is still possible, however, to achieve a rapid analysis:

(i) either by heating the column during the analysis in such a way as to accelerate the passage of the less volatile constituents (we have already seen that the logarithm of the retention volume is inversely proportional to the absolute temperature). This is chromatography with programmed temperature;

(ii) or by using several columns having different characteristics connected in series or parallel, each being responsible for the separation of a group of constituents with sufficiently similar properties. In this case a molecular filter (or activated charcoal) is generally used for separating the permanent gases (H_2, O_2, N_2, CH_4 and CO) **(18)(19)**, and silica gel for CO_2 and the hydrocarbons in C_2 and C_3 **(20)(21)(22)**. The separation of the light hydrocarbons in C_2 and C_3 can be carried out more quickly with activated alumina, and if it is also necessary to separate the heavier hydrocarbons, it is best to use a liquid solvent impregnated on an inert solid base. Among the more useful solvents we would mention dimethylformamide, propylene carbonate (both these very volatile solvents must be used at $0°C$), dimethylsulpholane, dimethylsulphoxide, hexamethylphosphoramide **(23)(24)**, etc. The porous and inert solid bases are either C_{22} brick (Chromosorb P) or celite 545 (Chromosorb W). In the so-called capillary columns ($\phi < 1$ mm) the walls of the column serve as a support.

Choice of detectors and carrier gases. The detectors are continuous analysers of binary mixtures (carrier gas + solute) responsible for recording the peaks, that is, for transmitting the concentration curves of each of the constituents at the output of the column.

They must have a very short response time and high sensitivity. Quite a wide range of these analysers has been produced commercially. Among them there are three in current use in industry **(26)(16)**.

Conductivity cell. This is the most widely used analyser because of its simple nature, reasonable price and acceptable sensitivity. It consists essentially of a wire heated by a stabilized electric current and placed in a cell through which the gas mixture to be analysed is passed. The equilibrium temperature of this wire, and therefore its resistance, with other things being equal (intensity of the heating current, flow and pressure of the gas, temperature of the gas and of the walls of the cell, etc.), depends on the thermal conductivity of the gas only; that is, if we are concerned with a mixture of two gases of different conductivities, on the concentration of one in the other.

In chromatography, the apparatus comprises two identical cells, one, through which the carrier gas passes before its entry into the column, serving as a reference, while the other, through which the mixture for analysis coming from the column passes, serves for measurement by means of a Wheatstone bridge arranged as shown in Fig. 6.5 (a). The change in the resistance of the hot wire in the measuring cell is converted into a potential difference.

It should be stressed that,

(*a*) in order to ensure the accuracy of the measurement it is necessary to work under perfectly stable and reproducible conditions: a stabilized current source for the hot wires, thermostatics regulation of the cell temperature, regulation of the flow and pressure of the gas, etc.;

(*b*) the apparatus will be all the more sensitive the greater the difference between the thermal conductivity of the carrier gas and the constituent in question. It can be seen from an inspection of Fig. 6.6, which gives the thermal conductivities of the usual gases at 0°C referred to that of air, that hydrogen and helium, because of their very high conductivity, are indispensable as carrier gases for the analysis of all other gases, and vice versa;

(*c*) the signal obtained is hardly ever a linear function of the concentration of the solute in the carrier gas over the whole range of concentrations **(14)**,

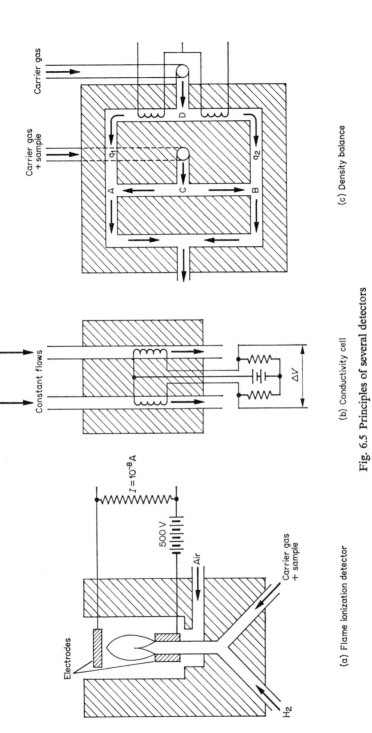

(a) Flame ionization detector

(b) Conductivity cell

(c) Density balance

Fig. 6.5 Principles of several detectors

but it can generally be assumed to a first approximation that this is the case over narrower limits. We would point out that there are irregular mixtures, hydrogen in helium is one of them, where the conductivity passes through a minimum at about 5% hydrogen. A mixture of this kind is obviously not suitable for analysis in the region of the minimum.

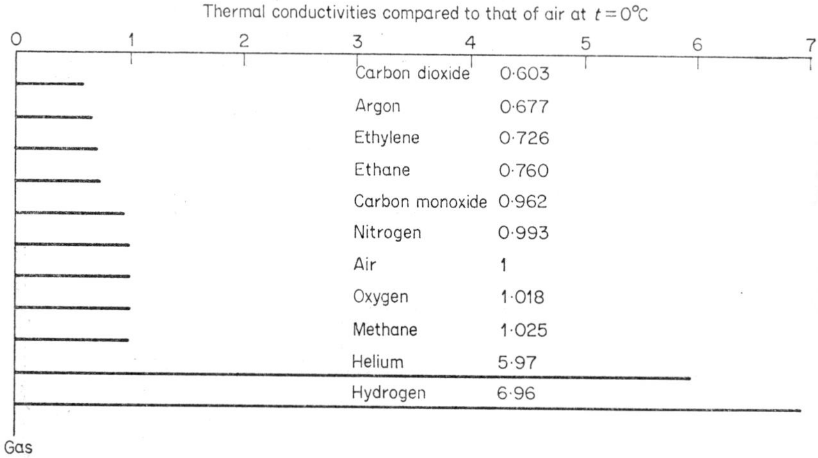

Thermal conductivities compared to that of air at $t = 0°C$

Gas	Value
Carbon dioxide	0·603
Argon	0·677
Ethylene	0·726
Ethane	0·760
Carbon monoxide	0·962
Nitrogen	0·993
Air	1
Oxygen	1·018
Methane	1·025
Helium	5·97
Hydrogen	6·96

Fig. 6.6 Thermal conductivity of several gases

The flame ionization detector: This detector makes use of the fact that the quantity of ions present in a hydrogen flame is modified by the introduction of another constituent. A diagram of the apparatus is given in Fig. 6.5 (b). The burner is supplied with a constant flow of hydrogen into which a flow (also constant) of the gas issuing from the chromatographic column is mixed, together with air (or sometimes pure oxygen) to support combustion. Two electrodes, one of which is placed above the flame and the other on the burner, produce an electric field in the flame, and collect the ion current. This current, which is very small, passes through a high resistance, and the voltage developed across its terminals is amplified and measured. Up to a certain limiting value, which depends on the arrangement of the electrodes, the signal remains a linear function of the concentration of the gas being analysed. The sensitivity is very good for substances which produce a large number of ions during combustion (hydrogen itself producing very few), that is, mainly hydrocarbons, of which a concentration of 1 ppm can be detected. The carrier is usually hydrogen but any other gas can be employed. This detector is much used for hydrocarbon analysis because of its very high sensitivity.

The density balance: As its name indicates, this apparatus detects the difference in density between the carrier gas and the constituent in question. Figure 6.5 (c) gives a vertical section of it; the distribution of the flows q_1 and q_2 of a reference gas entering at D, into the two circuits in the cell depends, other things being equal, on the pressure difference between points A and B, which is equal to the weight of the column of gas, that is, of the mixture being analysed which is admitted at C. The flows are measured by two hot wires arranged as a Wheatstone bridge. We would add that:

(*a*) the flows and the pressure of the gas, as well as the temperature of the cell must be kept quite constant during the measurement;

(*b*) the signal is an exact linear function of the concentration of the particular constituent in the carrier gas under suitable flow control conditions;

(*c*) the apparatus can analyse any constituent (with a density different to that of the carrier gas) however reactive it may be, because the hot wires are only in contact with the reference gas; and

(*d*) the sensitivity depends on the difference in density between the carrier gas and the constituent to be analysed. The light gases have the disadvantage of diffusing too quickly, which involves finding a very heavy carrier gas.

The three devices described above give a relative signal; they must, therefore, be calibrated for every substance it is desired to analyse quantitatively. We shall see later that this calibration is carried out on a chromatograph-detector assembly. It should be mentioned in conclusion that there are a number of other detectors less widely used (**16**): the mass spectrometer (mainly for identifying constituents), the infra-red spectrometer, the electron-capture detector, the radiation detector (**27**) and the mass detector (**28**).

Choice of a receiver

The concentration of solute in the carrier gas is converted into a potential difference which the receiver must measure and record. If this potential difference is a linear function of the concentration, the area under a peak corresponding to a constituent is proportional to the volume of this constituent introduced at the beginning of the column, whatever the shape of the peak, provided, of course, that there is no superposition, that is, that there is perfect separation.

The ideal receiver, therefore, will comprise a recording potentiometer which enables the quality of separation to be monitored, and an integrator which measures the area under the concentration peak representing the volume of the constituent. When the shape of the peak always remains the same, integration is no longer necessary, the height of the peak then being representative of its area.

Finally, if the potential difference is not a linear function of the concentration, a quantitative analysis based on the height or the area under the peak is only possible from a comparison with the results of a calibration, subject to the operating conditions during calibration and measurement remaining the same. In this case the measurement of the height of the peaks is sufficient.

We should add that the sensitivity (and therefore the accuracy, taking into account the reading error) of the integrators increases with the time of retention (longer peaks), whereas that of voltmeters, on the other hand, falls.

Recording voltmeters. These are required to have a low inertia compared with that of the detector, an adequate accuracy and a suitable scale of measurement (they are normally connected to the terminals of a manual or automatic attenuator). It is possible in this way to follow the profile of the peak and the stability of the base line (reference signal corresponding to the pure carrier gas), and to measure the height and the area under the peak (with a planimeter or by estimating), as well as the retention time.

Digital voltmeters. These do not provide for recording but enable the position of the zero and the height of the peak to be transmitted to a printer or a punch; they have the considerable advantage of retaining the same resolution over a very wide range of measurement.

Automatic integrators. These exist in various forms:

(*a*) Electromechanical integrators, which use ball and disc integration. The surface is converted into oscillations which are reproduced in the potentiometer recording. The price is relatively low but the accuracy is only moderate.

(*b*) Electronic integrators, which exist in two forms:

(i) analogue integrators; the signal from the detector, previously amplified, is applied to a standard integrating circuit. A differential circuit automatically controls the beginning and end of integration and the recording of the height and the area under the peak from the variations in the slope of the signal. Any drift in the base line can be compensated automatically by making the integrator operate between each peak for a given time. The result is converted into a potential difference.

(ii) digital integrators. These are similar devices to the above with the exception that the integrating circuit is replaced by a voltage (frequency) converter followed by an electronic pulse counter. The result displayed in digital form can be sent to a printer or a punch. The most up-to-date devices, equipped with an electronic clock, give the time of retention.

Calibration. The calibration relates to the whole unit (introduced volume + column + detector + receiver) operating under the measurement conditions. We proceed by introducing mixtures of known composition into the apparatus in order to record the characteristic area (or height) of the peak concentration in the volume of the constituent. If the separation is complete for a given constituent and the height of the signal is proportional to the concentration, the characteristic is a straight line passing through the origin, which is therefore defined by a single standard mixture. The measurement of the area under the peak then gives the exact volume required, whatever the shape of the peak; in other words the surface remains precisely the same if the control conditions of the chromatography have changed, with the exception, of course, of the sample volume which has been introduced. On this point we should make it clear that in nearly every apparatus, the procedure is to introduce the same volume of gas for analysis at atmospheric pressure. Hence, other things being equal, the signal will be proportional to this pressure, and should be corrected, therefore, if this varies. If the height of the signal is not strictly a linear function of the concentration the calibration must be carried out from point to point, and we have already seen in the previous paragraph that since the area no longer represents anything it is sufficient to measure the height of the peak. Finally, a number of cases are possible when the separation of the two constituents is not complete:

(*a*) If superposition has not taken place by the time maxima appear, the heights of the peaks are representative and we revert to the previous calibrations.

(*b*) If superposition is complete (no separation), which is the case for oxygen and argon in the molecular filter, corrections are possible only if we know the concentration of one of the constituents. Thus in our furnaces the concentration of argon which is proportional to that of nitrogen, is deducted from the nitrogen peak, and the height of the oxygen peak is corrected accordingly.

(*c*) The more complicated intermediate cases require interpretation of the chromatograms.

Practical realization

We have recently built a gas phase chromatograph in our laboratories which is intended to replace the Orsat instrument for analysing combustion gases. The improvements aimed at were as follows:

(i) more complete separation in respect of the following constituents: H_2, O_2, N_2, CH_4, CO, CO_2 and the principal hydrocarbons from C_2 to C_4;

(ii) an analysis time (10 minutes) which would be short enough to enable the two Orsat analysers in use to be replaced during our tests;

(iii) improved accuracy (better than $\pm 0\cdot 2\%$ absolute) for all constituents, particularly H_2, CH_4 and N_2, and certainly much higher reproducibility of the results (1% relative);

(iv) automatic operation after the specimen flask has been placed in position;

(v) the possibility of modifying the apparatus quickly so as to adapt it to other uses, such as the analysis of tracer gases, search for traces, more detailed analysis of hydrocarbon vapours, etc.

Description. These requirements are met by the equipment shown diagrammatically in Fig. 6.7 which comprises three chromatographic circuits:

(i) a circuit for analysing hydrogen with nitrogen as the carrier and a 5A molecular filter column 90 cm long and of 4 mm internal diameter;

(ii) a circuit for analysing the permanent gases, O_2, N_2, CH_4 and CO with helium as the carrier, and a molecular filter column with the same characteristics as above. It should be noted that oxygen and argon are not separated and that this must be taken into account for calibration;

(iii) a circuit for analysing CO_2 and the hydrocarbons with helium as the carrier and two parallel columns which can be changed over: a column of silicagel 25 cm long and 4 mm internal diameter, and either a column of activated alumina (1·10 m long, 4 mm diameter) or a column 6 to 8 metres long, 4 mm in diameter, filled with C_{22} brick impregnated with 30% by weight of hexamethylphosphoramide for separating hydrocarbons.

Each circuit comprises:

(i) a supply system for the carrier gas providing constancy of the pressure at the head of the column by two pressure regulators placed in series, the second being a precision Negretti–Zambra pressure reducer, and of the flow due to a charge regulator in a control needle valve (Fig. 6.8);

(ii) a conductivity cell with a supply stabilized at 200 and 250 mA for the hot wires made by Becker (Netherlands);

(iii) a six-way sampling valve, Aerograph (U.S.A.) type with electromagnetic control which enables a constant volume of gas for analysis to be injected into the current of carrier gas (Fig. 6.8);

(iv) one (or two) columns of copper (or stainless steel) with a 4 mm internal, and a 6 mm external diameter, following a cylindrical spiral, 15 cm in diameter;

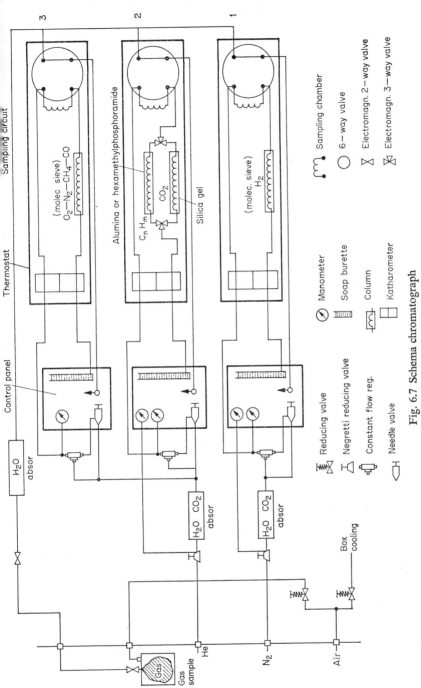

Fig. 6.7 Schema chromatograph

1. Entry of carrier gas
2. Entry of sample
V Sample volume

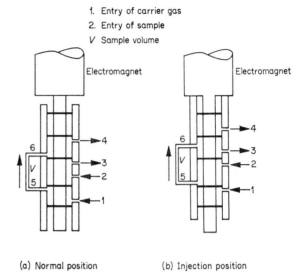

(a) Normal position   (b) Injection position

Calibration valve

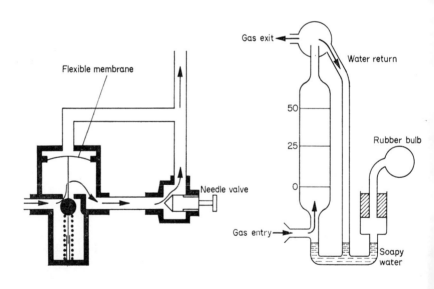

Flow regulator   Soap bubble flow meter

Fig. 6.8 Auxiliary equipment

(v) an enclosure, thermostatically controlled to 0·1°C between 20°C and 250°C, built by Becker, in which the column or columns, the conductivity cell and the sampling valve are located;

(vi) monitoring and control equipment; pressure gauges, soap bubble flow meters (Fig. 6.8), various valves, etc.

The whole assembly is installed in a metal cabinet which is itself thermostatically controlled to about 1°C, so as to eliminate the effects of ambient temperature fluctuations.

Functioning of the instrument. With the carrier gas currents stabilized, the electrical supply is connected to the thermostats, the receiver and the conductivity cells (in the absence of a gas current, the hot wires may be damaged). When the temperatures are stabilized, the flows are adjusted by means of the needle valves and the flow meters. The flask containing the gas for analysis is placed on its stand and connected to the tube T_1 (Fig. 6.7) leading to the sampling valves. The compressed air is then connected to the drain tube T_2 of the gas-tight box, and the cock R_1 is opened. A disc programmer and numerous relays then control automatically all the remaining operations in the following order for each chromatographic circuit: the valve V_1 is open and the 6-way sampling valve opened (Fig. 6.8) in such a way that the gas for analysis, propelled by the compressed air, fills the sampling loop and exhausts to free air. After a time sufficient for purging the line, the valve V_1 is closed in order to establish atmospheric pressure in the sampling loop. The sampling valve is then re-set to its normal position and the carrier gas forces the gas to be analysed, which is contained in the sampling loop, towards the column. In this way a definite volume of air is introduced, always under the same conditions (the operations are carried out automatically) at atmospheric pressure. The attenuation to be applied to each peak in order that it shall fall within the scale of the recorder, taking into account the possible limits of concentration, is also programmed automatically. The instructions for the three chromatographic circuits are combined as shown in Table 4, in such a way that the analysis of the six gases H_2, CO_2, O_2, N_2, CH_4 and CO is carried out in 4 minutes under completely automatic conditions; pilot lights enable the course of the analysis to be followed on a panel on which the three chromatographic circuits are shown diagrammatically.

When hydrogen or methane, signs of the possible presence of hydrocarbons, have been detected, the analysis can be prolonged by manual control for the hydrocarbons on the column of alumina (or of hexamethylphosphoramide), as long as necessary: C_2H_6, C_2H_4, C_3H_8 and

C_3H_6 will appear in 3 minutes and the first hydrocarbons with C_4 in 5 minutes. The change-over of the CO_2 and C_nH_m columns is carried out by means of a 6-way valve, controlled pneumatically. The receiver is an electronic recording potentiometer, with a 0·25 s response time over a full scale of 0 to 1 mV; the possible attenuations are 2, 5, 10, 20, 50, and 100.

Function variables. (i) The molecular filter is prepared in accordance with the method recommended by Farrerius and Guiochon (**21**): the fine particles are eliminated by elutriation under a counter-current of water (or by decantation) so as to retain only a very restricted granulometry (315 μ–400 μ). The water is then eliminated by passing through a crystallizing pan containing a dehydrating agent, and then by heating to 80°C for three hours plus 250°C for 12 hours in a stove under vacuum. Activation is carried out by heating to 400°C for 12 hours under a current of dry air; (ii) the silica gel, also filtered to between 200 μ and 250 μ, is activated by heating to 200°C for 24 hours under a current of dry nitrogen; (iii) the alumina (granulometry 200 μ) is activated at 400°C for 12 hours under a current of dry air; (iv) the other column for separating hydrocarbons is prepared with 70 g of C_{22} brick (315 μ–400 μ) to which a solution of 30 g of hexamethylphosphoramide dissolved in ether is added. This is thoroughly mixed before allowing the ether to evaporate into the air and then in the column with the stream of carrier gas. Since the hexamethylphosphoramide is quite volatile, it is advisable not to exceed 55°C during operation because of the risk of reducing the life of the column considerably; (v) the operating temperatures (between 35°C and 45°C) and the flows of carrier gas (of 3 to 6 1/mn) are adjusted for each column so as to preserve adequate separation of the peaks for a minimal analysis time. Experience shows that it is very difficult to build two identical columns which are interchangeable so that with each new column it is necessary, in order to regain the operating conditions of the previous column, to adjust the flow and the temperature. This operation is possible on our installation which comprises three thermostats and three flow regulators: in this way we avoid revising the programming each time; (vi) the sampling volume (at present 0·1 cm^3 in the H_2 circuit, 1 cm^3 in the CO_2 + hydrocarbon circuit, and 0·25 cm^3 in the permanent gas circuit) is taken small enough for the signal (peak height) given by the conductivity cells to remain proportional to the concentration within the limits defined in Table 1, for all the constituents analysed. The calibration is greatly facilitated in this way. We should add that the sampling volume, which is that of the 6-way valve loop shown in Fig. 6.8, always remains the same but that the pressure of the gas to be analysed equals the atmospheric pressure at the time of injection (see Functioning of the Instrument on page 141). This pressure should be recorded, therefore, at the moment of measurement, for subsequent corrections;

(vii) in order to prevent the separating power of the columns from falling too quickly with time:

(*a*) the carrier gases are purified by cartridges which absorb the carbon dioxide and the water vapour (KOH + P_2O_5 in granules);

(*b*) the gas to be analysed is freed from water vapour by a P_2O_5 cartridge before arriving in the 6-way sampling valve;

(*c*) the molecular filter columns are preceded by a short and easily interchangeable column (10 cm), containing activated alumina for the irreversible adsorption of carbon dioxide, and MnO_2 for the adsorption of SO_2 and certain hydrocarbons.

Performance. The advantages sought for are in fact achieved by this apparatus.

(i) the analysis of the major constituents, H_2, CO_2, O_2, N_2, CH_4 and CO, takes only 4 minutes and is carried out automatically (Fig. 6.9);

(ii) experience has shown that even in fuel-oil flames relatively few light hydrocarbons are found, and these only rarely, and when there are any it is almost entirely in the form of a mixture of ethane, ethylene, and propylene, a mixture which can be made to appear on the alumina column in 3 minutes (Fig. 6.10). The average duration of the analyses remains, therefore, in the region of 4 minutes;

(iii) the separation of all these constituents is good except for:

(*a*) the separation of methane and the trace of the nitrogen peak when the molecular filter column is part-used;

(*b*) the separation of ethane and ethylene which is not possible with hexamethylphosphoramide. On the other hand, this can be done with alumina (or silica gel);

(*c*) the separation of propane and the trace of the CO_2 peak with the hexamethylphosphoramide column (Fig. 6.11). We also find that with a 4 minute period the end of the CO peak is not recorded, or the beginning of the CO_2 peak. This is not important since we are only measuring the height of the peak for quantitative purposes. We have verified, moreover, that so long as the analysis is always carried out under the same conditions, it is not necessary to measure the height of the peak above the base line, and that it is sufficient to take the difference of level between two characteristics of the peak: the summit and a point fixed by some means (for example, a change in sensitivity). Subject to the condition, naturally, that there is no interference with any other peak, this difference in level is

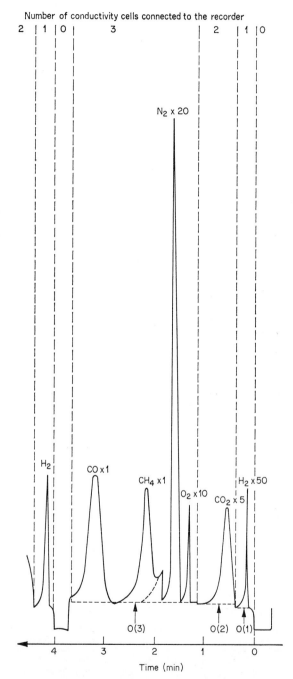

Fig. 6.9 Analysis of the permanent gases

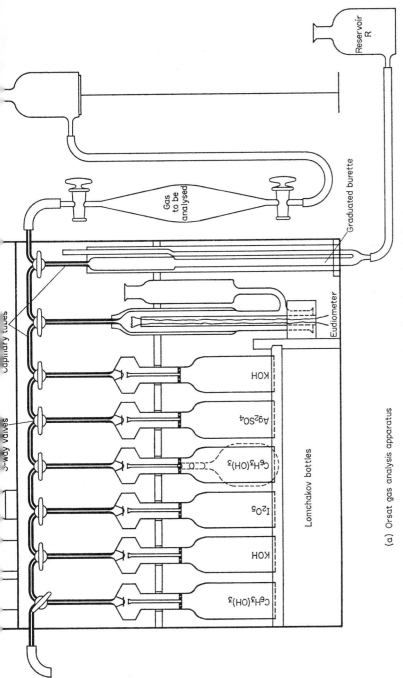

(a) Orsat gas analysis apparatus

Fig. 6.10 Separation of hydrocarbons by activated alumina

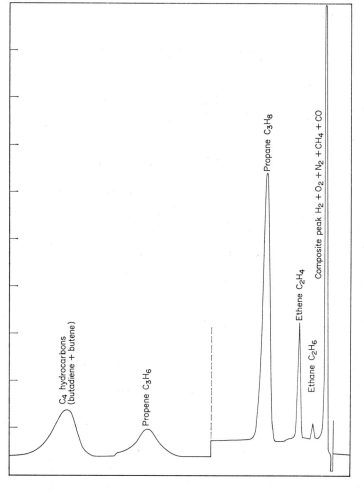

Fig. 6.10b

strictly proportional to the height of the peak; that is, to the concentration of the gas;

(iv) the use of two carrier gases, nitrogen and helium, has made it possible to obtain high sensitivity for all the constituents analysed with the conductivity cell $(2 - 1 - 3)$ (the hydrocarbons could be analysed more accurately with a flame ionisation detector but at the price of too great a complexity of the instrument). The ranges of measurements specified in Table 6.1 fall within the full scale of the recording potentiometer (0–1 mV) with the following attenuations programmed in advance:

Gas	Column	Volume of sample in cm³	Attenuation
H_2	1	0·1	50
CO_2	2	1·0	5
O_2 N_2 CH_4 CO	3	0·25	10 20 2 2
Hydrocarbons	4	1	5

The reproducibility of the measurements in all cases is better than 1 mm of paper over the whole scale which, taking the attenuations into account, corresponds to 0·4% of N_2, 0·1% of O_2, CO_2, CO, CH_4 and 0·05% of H_2, $C_2H_4 + C_2H_6$, etc.

This reproducibility which can be estimated as better than 1% relative when the attenuation is adjusted manually as a function of the effective concentrations in the sample analysed is obviously due to:

(*a*) the double thermostat control apparatus which makes it possible to operate always under the same temperature conditions to about 0·1°C for the column, the detector and the sampling, and to 1°C approximately for the other auxiliaries;

(*b*) the regulation of the carrier gas flow;

(*c*) the protection of the columns (in 1000 analyses the calibration constants have not varied more than a few per cent);

(*d*) corrections to the atmospheric pressure (see Calibration on page 137).

(v) the minimum detectable quantities under automatic operating conditions, and taking into account the instability of the base line, amounts at the most to 0·05% approximately;

(vi) finally the absolute precision depends essentially on the accuracy with which the sample mixtures are realized. When a number of sample mixtures are used which are known to about 0·1 % absolute, it is necessary to

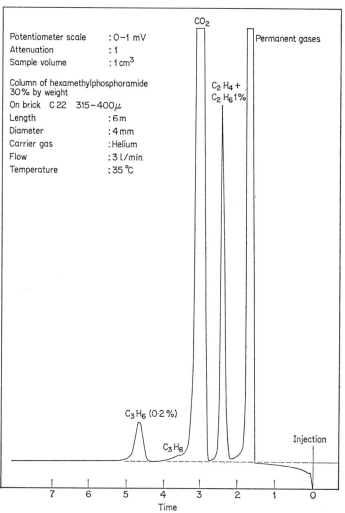

Fig. 6.11 Separation of hydrocarbons by hexamethylphosphoramide

add this value to the deficiency in reproducibility in order to obtain the absolute value. In our case it is at the end of the scale, 0·5 % for N_2, 0·2 % for O_2, CO_2, CO, CH_4; 0·15 % for H_2, C_2H_4, C_2H_6, C_3H_6, etc;

(vii) figure 6.12 gives the characteristics height of peak-volume concentration for the 6 permanent gases. It should be noted that these characteristics

	Curve No.	Nature of gas	Attenuation of signal
o	1	H_2	20
●	2	CH_4	1
△	3	CO	1
▲	4	H_2	50
□	5	O_2+0·95% dA	10
■	6	CO_2	5
+	7	N_2	20
✳	Oxygen + variable % of argon		

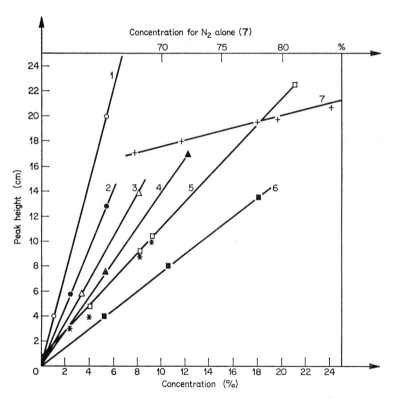

Fig. 6.12 Peak height volumetric concentration characteristics for the principal gases analysed

6

are effectively straight lines and that the dispersion of the calibration points is in conformity with the precision required.

Conclusion

Chromatography, which offers numerous possibilities in the domain of discontinuous analyses seems particularly suited to the separation of very complex mixtures. It calls for relatively uncomplicated equipment and gives an excellent reproducibility and very good measurement accuracy. It is especially valuable for automatic operation.

6.3 Equipment for continuous analysis

There are many types of continuous analysers available commercially, based on physical principles, which have the property of being more or less selective for certain gases, or categories of gases. It is possible to mount these either in series or branched off from the sampling circuit, and in this way to make a continuous analysis of a mixture of several gases. We use an arrangement of this kind for monitoring the composition of flue gases with the following two devices.

Paramagnetic analyser

This makes use of a property which only a few gases possess (oxygen, nitric oxide NO, nitric peroxide NO_2) of being paramagnetic (the magnetic susceptibility X, is positive and a function of temperature in accordance with Curie's Law $X = C/T$, where C is a constant and T is the absolute temperature) while all the other gases are diamagnetic ($X < 0$ and is independent of temperature). There are different versions of the apparatus which all operate in the following way: the paramagnetic gas of the mixture to be analysed is drawn into the field of a magnet from which it is expelled by heating, X falling with temperature (Fig. 6.13). Measurements are made of the resistance of hot wires placed in the current of gas formed in this way (called the magnetic wind), the intensity of which is proportional to the concentration of paramagnetic gas. In practice many other factors also affect the temperature of the hot wires: the temperature, pressure and flow of the gas in the apparatus, the composition (viscosity and thermal conductivity) of the gas mixture to be analysed, the stability of the electrical supply, etc. This is why paramagnetic analysers must incorporate (14):

(i) an effective thermostatic control of the measurement chamber;
(ii) a stabilized supply for the hot wires;
(iii) a regulation system for the flow and the pressure of the gas.

As a rule the regulated flow exhausts at atmospheric pressure and certain corrections are made as a function of the latter.

There remains only the elimination of the effect of the composition of the gas mixture to be analysed. It is quite obvious in the first place that the analyser can only be used when the mixture contains a paramagnetic gas (in our case oxygen). Again it is advisable to analyse gases which have been dried beforehand in order to avoid condensation inside the apparatus on the one hand, and also because water vapour has noticeably different physical characteristics to those of the majority of other gases. Finally

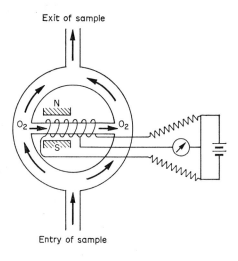

Fig. 6.13 Principle of the paramagnetic analyser

the errors will be kept small if the conditions of measurement and calibration are very similar. Otherwise the corrections are practically impossible to calculate, and the calibration must be done again. We should mention, however, that there is at least one apparatus (Siemens, West Germany) in which the effects of variations in the gas to be analysed are reduced considerably by means of a symmetrical arrangement of four measurement chambers; only two of these have the magnetic wind passed through them, the other two are for reference.

In all cases the resulting error must not exceed a few per cent of the maximum scale value, the range of measurement being selected by the user. The response time for 90% of the signal is of the order of 10 seconds, including the delay due to the time the gas takes to arrive in the measurement chamber.

Infra-red analyser

This makes use of the property possessed by the majority of compound gases of selectively absorbing certain long-wave bands of infra-red radiation. The principle of the apparatus is illustrated in Fig. 6.14. The radiation emitted by an infra-red source, and filtered if necessary, traverses the measurement chamber containing the gas to be analysed, and undergoes absorption there according to the wavelength bands characteristic of this

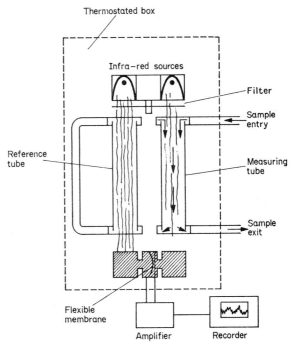

Fig. 6.14 Principle of infra-red analyses

gas, in proportion to the product of the thickness of the chamber and the partial pressure of the gas. The residual radiation is then absorbed by a receiving cell filled with the constituent to be analysed, which heats up, increases in pressure and causes the displacement of a pressure diaphragm, this displacement being measured by a variable capacitance transducer. The signal is proportional to the concentration sought for. Several points arise in this connection (**14**).

(i) Only the wavelengths absorbed by the constituent to be analysed are taken into account by the receiving cell which is filled with the same gas. The apparatus is selective, therefore, although the source is not monochromatic, except when the mixture introduced into the measurement cell contains other constituents whose absorption band interfere with those of the gas being analysed. In this case it is necessary to filter the radiation emitted by the source by placing in front of the measurement tube another tube containing the parasitic gases; that is, by holding back the radiation emitted in the common wavelength bands. Compensation of this kind is not possible for particles (grey bodies which absorb all wavelengths) or for water vapour (condensation, complex spectrum), so that the gases must be carefully dried and freed from dust before introducing into the analyser.

(ii) In order to compensate for the effects of variable ambient conditions, particularly temperature, the commercial devices incorporate two identical tubes of which one filled with non-absorbing gas (N_2) serves as a reference. The two receiving cells are built as one unit and the pressures in them act in opposition across the diaphragm.

(iii) The measurement ranges vary from 0%–$0\cdot01\%$ to 0%–100% according to the thickness of the measurement tubes being used, these tubes being interchangeable. When using them it is necessary to ensure that the product of the pressure and the thickness is constant; for this purpose the gas is introduced into the tube through an annulus and also at a low rate which gives the desired response time for avoiding over-pressure. In addition, it is best to operate by discharging the gas into the apparatus with exhaust to atmosphere so that the pressure inside it is very near that of the atmosphere. The correction factor to be applied to the result is then the ratio of the pressure at the time of measurement to that at the time of calibration.

(iv) The commercial devices are thermostatically controlled and supplied with voltage and frequency stabilizers. The possible errors do not exceed a few $\%$ at the top of the scale, and the response time, which depends on the flow of gas fed into the apparatus, can be less than one second. The gases which can be analysed cover nearly all the compound gases; CO, CO_2, CH_4, the oxide of nitrogen, hydrocarbons, etc.

Other methods

There are other types of analysers which are based on mass, viscosity, thermal conductivity, etc., some of which have already been described in the section on Choice of a Receiver on page 135. By making use of certain corrections which allow for the complexity of the gas mixture containing the constituent to be analysed, it is possible to use any of these devices. For example, the conductivity cell is often employed to measure the concentration of hydrogen in a gas mixture of which the conductivity is very close to that of air (20).

6.4 Mass spectrometer

This instrument, which is considered even today as an awkward laboratory device, complicated in design, difficult to set up, and very costly, has so far found relatively few applications in traditional industry, despite its remarkable performance. It appears, however, that because of the progress in electronics and vacuum technology, makers today are ready to market industrially orientated versions. We have made a study in our laboratories of the possibilities of applying it to the examination of industrial diffusion flames (**13**).

Principle (Fig. 6.15)

The gas mixture to be analysed, introduced into the apparatus at very low pressure (10^{-5} torr), is ionized by a current of electrons. The ions formed in this way, when accelerated by an electric field, acquire a kinetic energy of $\frac{1}{2}mV^2 = Ue$ (*e* being the charge on an electron, *U* the accelerating potential, *V* the velocity of the ion, and *m* its mass, which is practically that of the original atom), and are then subjected to the action of a magnetic field *H* normal to their trajectory. The ions then describe circles of radius *R* such that the centrifugal force balances the electromagnetic force $HeV = mV^2/R$

$$R = \frac{1}{H} \times \left(2U\frac{m}{e}\right)^{\frac{1}{2}}$$

In other words, there are as many trajectories as there are ratios m/e. Ions of different mass are separated and with a suitable geometrical arrangement, ions corresponding to the same m/e are focused onto a collector, which is connected to a direct current amplifier, the output signal of which is proportional to the intensity of the ion current received. With the collector fixed, a progressive modification of the magnetic field *H* makes it possible to collect successively the ions corresponding to the various atomic masses of the constituents introduced, that is, to read on a receiving recorder a trace of the successive peaks called the mass spectrum, the abscissa being m/e and the ordinate the corresponding magnitude (height of the peaks) of the ion current. In practice, in order to obtain an adequate and stable current of ions it is necessary to use quite high energy electrons (70 eV). Under these conditions the molecules may be disintegrated, so that the spectrum of a pure substance does not show a single peak but a peak without a break called a parent peak, accompanied by peaks corresponding to possible fragments ionized once or several times (Fig. 6.16). The mass spectrum is therefore more complicated than one

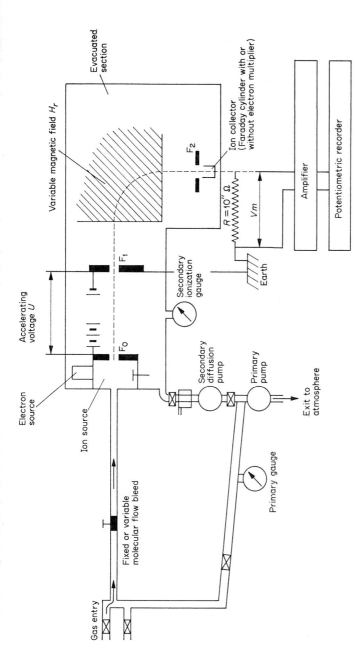

Fig. 6.15 Principle of the mass spectrometer

would think although this complication has, however, the advantage of allowing a distinction to be drawn between two substances of the same molecular mass but of different chemical formulae.

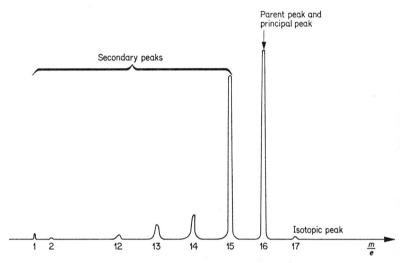

Fig. 6.16 Mass spectrum of pure methane

Practical application

Figure 6.15 shows the basic arrangement of mass spectrometers of classical type which comprises:

(i) a device for introducing the gas into the analyser which can be either discontinuous or continuous (Fig. 6.17). This device is intended to transform the gas to be analysed from atmospheric pressure to the very low pressure (10^{-5} torr) maintained in the apparatus and to ensure the constancy of the flow introduced in order to permit a quantitative analysis. For a discontinuous analysis a small sample volume, perfectly reproducible, is released into a flask of greater volume connected to the apparatus by a molecular flow bleed with a fixed or adjustable opening. For a continuous analysis, the mixture to be analysed is drawn in through a series of valves or a capillary tube controlled by a needle valve, by an auxiliary pump which can maintain a pressure of 10^{-2} torr, the junction with the molecular flow being secured where the pressure is sufficiently low. We recall that in molecular flow the flow of a gas is proportional to its partial pressure and independent of the partial pressure of other constituents. On the other hand, other things being equal, the light gases flow more quickly than the heavy gases;

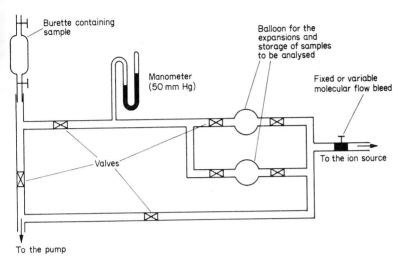

(a) Discontinuous introduction

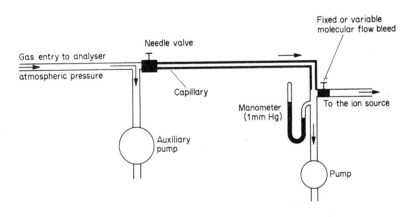

(b) Continuous introduction

Fig. 6.17 Equipment for introducing gas into the ion source

(ii) a source of ions which ionizes the gases which are introduced, generally (Bleakney source) by a regulated current of electrons emitted by
a heated filament, and then accelerates the ions formed in this way in a
stabilized electric field, and finally produces a beam of ions which has
negligible dispersion;

(iii) a tube in which the beam of ions is propagated under the influence of a magnetic field;

(iv) a variable-field electro-magnet which enables the mass spectrum to be explored;

(v) an ion collector which is a Faraday cylinder (it is possible to precede this by an electron multiplier in order to improve the sensitivity from 10^3 to 10^6 times). In order to analyse several constituents simultaneously some instruments have several collectors placed side-by-side, the magnetic field being held constant;

(vi) a direct current pre-amplifier and main amplifier for the ion current picked up on the collector;

(vii) a tube pumping device comprising two pumps placed in series; a primary vane pump (10^{-3} torr) and a diffusion backing pump (10^{-7} torr). The vacuum in the apparatus must be high (5×10^{-6} torr) in order to limit the deceleration of the ion beam. We note that the diffusion pumps operate either with mercury, the vapours of which must be collected in a liquid nitrogen trap, or with low vapour pressure oil which does not necessitate a trap but is less efficient;

(viii) an ionization gauge for controlling the vacuum in the tube;

(ix) stabilized supplies for the electro-magnet and the source, the monitoring and control apparatus, safety systems, etc.

Performance

We have carried out some tests on a small SM 100A apparatus built by the C.S.F. (France) which enables all gas compounds of atomic mass less than 100 to be analysed at 150°C.

Stability. This can be affected at different levels;

(i) amplifier: with our recording potentiometer which has a response time of 0·25 second, the amplitude of the background noise does not exceed ±0·1 mV which in the case of nitrogen, at peak 28, represents a volume concentration of less than 0·01 % (100 ppm). On the other hand the zero drift may be much larger if care is not taken to operate under favourable conditions (stable supplies, good earthing, constant temperature, etc.) which are sometimes difficult to obtain in a factory;

(ii) ion source: the ionizing electron beam is controlled by the current trap which reaches the anode and the energy of the electrons can be regulated at will from 10 to 20 eV with an accuracy of ±0·05 eV. By this means good reproducibility of the ion beam is determined under the usual operating conditions, that is, at 70 eV. The accelerating voltage is also stabilized, but distortion of the electric field resulting from contamination of the source by the deposition of tungsten carbide and various oxides is the cause of a slight but continued drift of the signal, a drift which can be limited by periodic cleaning (a few hours work every month);

(iii) the system of introduction: we have seen that this consists essentially of a fixed or adjustable molecular flow bleed, with measurement of the pressure up- and down-stream. It is necessary, of course, to keep the geometrical characteristics of the bleed perfectly constant during the measurements. This implies that there is thermostatic control of the equipment and complete filtration of the gases before the analysis in order to prevent any contamination. Thus the monitoring and regulation of the flow is carried out by measuring the pressure up- and down-stream. The pressure down stream is so low that it has no effect on the flow, and merely expresses its magnitude. It is measured by an ionization gauge the reading of which is the sum of two terms:

(*a*) the basic pressure of the system which results from the degassing of the wall, the vapour pressure of the diffusion pump fluid, etc.;

(*b*) the inlet pressure which is itself proportional to the sum of the partial pressures of the constituents, taking into account the sensitivity factors $N_2 = 1.07$; $O_2 = 0.85$; $H_2 = 0.58$; $H_e = 0.19$; water vapour $= 1$, etc. referred to dry air; so that the gauge can only give an order of magnitude which is very useful for controlling the instrument, but which cannot be employed for precise corrections. We recall that term (*b*) depends on the bleed opening, the pressure up-stream, and also on the composition of the mixture, the light gases passing more quickly than the heavier in molecular flow (a flow inversely proportional to the square root of the density).

Finally, the regulation of the input flow is only possible for a given bleed opening by measuring the pressure upstream, the magnitude of which is of the order of millimeters of mercury. The accuracy of the quantitative analysis will be a function of that of the manometer used. For this reason the absolute liquid column manometers (mercury or oil) are not suitable, and it is necessary to use diaphragm manometric transducers, an electrical displacement transmitter based on variations of mutual inductance or capacitance. The scale of these transducers should be a few millibars absolute with the ability to withstand large overloads; that is, atmospheric pressure (it should be noted that the tube vacuum can serve as a reference for a differential type of pick-up). The height of the peak depends linearly on the partial pressure of the corresponding gas in the source, that is, in fact, on its partial pressure upstream of the fixed bleed. This property of spectrometers makes it possible to calibrate by the analysis of pure gases only, and simplifies the scrutiny of the measurements as well as possible corrections for variations in the pressure upstream of the inlet.

Sensitivity and response time. The height of the principal peak (the highest peak of the spectrum which is usually the parent peak) for each of the principal gases present in our furnaces, under average control conditions reaches about 2 volts for 100%, which indicates that a millivolt represents a volume concentration of 0·05%. Taking into account the background noise, one can easily measure to 0·01% absolute (100 ppm). In the case of hydrogen and helium, which are difficult to ionize, the sensitivity is 2 to 3 times less, although the admission is greater for the same upstream pressure. A precision of this kind is quite sufficient in industry except for the detection of traces. In this case it is necessary to add an electron multiplier which increases the sensitivity to a considerable extent (10^3 to 10^6 times) but which has the drawback of not being very stable. As to the response time of the spectrometer itself, this is practically that of the chain of electrical measurement following the collector, and which comprises a pre-amplifier, an amplifier and a receiver. It is possible to obtain response times of less than 0·1 second, which is a very considerable advantage in many applications where the apparatus operates continuously. A different approach is that of continuing the analysis by scanning the mass spectrum, which depends on the range of the useful spectrum, the possibility of having to change the sensitivity by hand, and on the rate of scanning being compatible with the recording medium. With the SM 100A it takes several minutes to go from mass 12 to mass 100. It is possible to gain time by programming on the peaks required for the measurement, but the best solution is still that of continuous operation with a fixed magnetic field and several collectors: unfortunately the number of collectors is limited to 5 for reasons of size.

Resolution and interference. The resolving power expresses the ability of the apparatus to separate the peaks corresponding to two neighbouring masses. For the analyses with which we are concerned, this is usually quite adequate. On the other hand, the superposition of peaks of the same mass produced by different substances is extremely annoying because it complicates the examination of the results considerably (for example mass 28 is given, among others, by carbon monoxide, nitrogen, ethane and ethylene). Separation is then affected by referring to the secondary peaks; the quantitative analysis of n identified constituents (or assumed to be present) amounts to the solution of a linear system of n equations:

$$h_1 = A_{11}X_1 + A_{21}X_2 + \ldots + A_{i1}X_i + \ldots + A_{n1}X_n$$

$$h_i = A_{1i}X_1 + A_{2i}X_2 + \ldots + A_{ii}X_i + \ldots + A_{ni}X_n$$

$$h_n = A_{1n}X_1 + A_{2n}X_2 + \ldots + A_{in}X_i + \ldots + A_{nn}X_n$$

where: h_i represents the height of the peak $m/e = i$;

A_{ij} represents the amount in the peak $m/e = i$ of the constituent j (value obtained by calibration);

X_i represents the volume concentration (or partial pressure) of the constituent i.

Since the apparatus is never perfectly stable, the coefficients A_{ij} drift continually, and the calibration must be repeated quite frequently. As a result, the evaluation of the inverse matrix of the system cannot be made once and for all.

We recall that quantitative analysis is only possible if there is no interference, or at least very little, as is the case for monitoring gases at the end of combustion (more generally for a mixture of so-called permanent gases H_2, O_2, N_2, CH_4, CO_2, CO, SO_2) or if suitable means of calculation are available.

Another difficulty, equally serious, arises from the existence of a residual spectrum due basically to the adsorption of certain gases in appreciable quantities by the walls of the instrument, which is the case for water vapour, hydrosulphuric acid, etc., gases which cannot be analysed quantitatively with precision.

Conclusion

While the mass spectrometer has two major inconveniences—interference and the residual spectrum—it has, on the other hand, a great many advantages over other methods of analysis: simplicity of calibration which gives the measurements an absolute character, the ability to analyse continuously and simultaneously a number of constituents, absence of effects (apart from interference between masses) due to the complexity of the mixture being analysed, rapid response, negligible power consumption, etc.

6.5 Water vapour analysis

The problem of the quantitative analysis of water vapour is very specific, since it calls for the use of special probes and analysers. We have already seen, in fact, that the analysers described previously do not give the water vapour concentration and that the suction line must be heated to 150°C in order to avoid any condensation of acid vapours formed with SO_3 and SO_2.

There is a great variety of hygrometers on the market which measure either the dew point (20) or the weight of water vapour (retained by absorption or condensation) or the electrical conductivity of a salt in hygrometric equilibrium with the gas. These devices, which are all discontinuous in operation, are hardly suitable in our furnaces both because the presence of traces of sulphide involves large errors and because the scales of measurement are in general too small. There is, however, a practical interest in dew-point hygrometers capable of giving the temperature of condensation of acid vapours which play an important part in corrosion.

We would also point out that the simultaneous analysis of SO_2 and water vapour is possible by chromatography.

Despite several attempts, we have so far not perfected a method which permits accurate continuous analysis of the concentration of water vapour in the presence of traces of sulphide. Certain discontinuous methods are feasible using reactants which absorb water vapour selectively, but these are much more appropriate to the laboratory than to industrial practice.

Chapter 7

Analysis of Solids

Two types of analysis are used for the sampled solids—chemical and granulometric. These are carried out by means of well-known commercial instruments, with clearly defined instructions for their use, of which we recall here only the main features.

7.1 Chemical analysis

We have available two analysers: the first for a rapid determination of the ash content (or of unburnt material), and the second for an elementary analysis.

Rapid determination of the ash content

This is done by a Pozetto type apparatus, a diagram of which is shown in Fig. 7.1. The solid to be analysed is weighed dry in a small refractory cupel which is to be introduced into a furnace heated electrically to about 700°C, where the combustion takes place under a current of pure oxygen. Some preliminary tests enable the measurement conditions to be fixed: the cupel charge, the oxygen flow (about 10 l/min measured across a restriction), and the duration of the analysis (5 to 10 minutes). After combustion of the unburnt residues, the cupel is withdrawn, placed in a crystallizing pan for cooling, and reweighed. The weight of the ash content is obviously the weight of the ashes remaining from the weight of original solid.

Elementary analysis

A Heraeus apparatus, illustrated in Fig. 7.2, is used for this purpose: a very small quantity of the solid or liquid to be analysed is weighed in a small crucible which is then introduced into a quartz tube through which various pure gases can be passed as required, and which can be raised to different temperature levels.

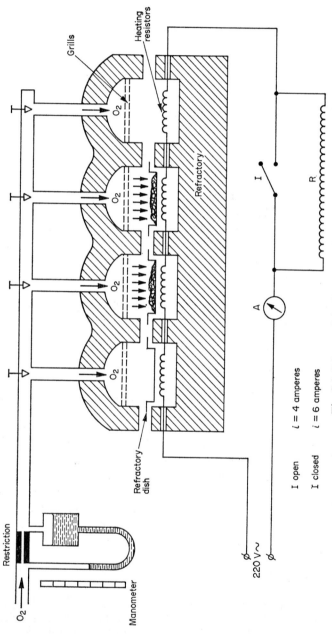

Fig. 7.1 Scheme of the Pozetto analyser

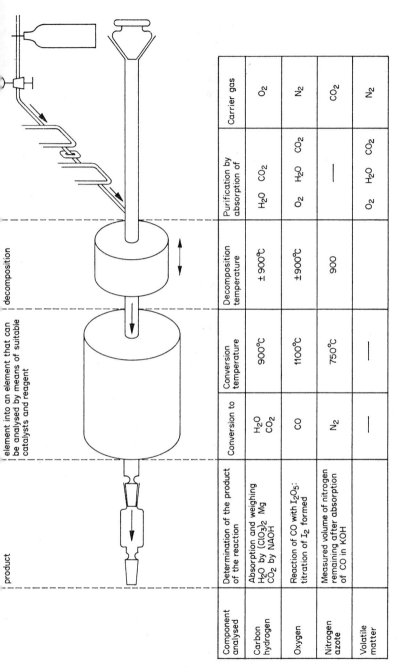

Component analysed	Determination of the product of the reaction	Conversion to	Conversion temperature	Decomposition temperature	Purification by absorption of	Carrier gas
Carbon hydrogen	Absorption and weighing H_2O by $(ClO_3)_2$ Mg CO_2 by NAOH	H_2O CO_2	900°C	±900°C	H_2O CO_2	O_2
Oxygen	Reaction of CO with I_2O_5: titration of I_2 formed	CO	1100°C	±900°C	O_2 H_2O CO_2	N_2
Nitrogen azote	Measured volume of nitrogen remaining after absorption of CO in KOH	N_2	750°C	900	—	CO_2
Volatile matter		—	—		O_2 H_2O CO_2	N_2

product

element into an element that can be analysed by means of suitable catalysts and reagent

decomposition

Fig. 7.2 Principle of the Heraeus apparatus

In this way a selective reaction can be produced, the gaseous products of which are drawn along by the carrier gas, treated in a furnace, and absorbed and weighed in the laboratory after passing through the quartz tube. Figure 7.2 shows the procedure to be adopted for the elementary analysis of carbon.

7.2 Granulometric analysis

In our laboratories these analysis are concerned with two quite different powders:

(i) solid particles sampled in flames of pulverized coal and whose characteristics range from those of the pulverized coal fed into our burners (Rosin–Rammler distribution, $1\ \mu < \phi < 250\ \mu$) to those of the fly-ash evacuated in the flue (ϕ from $100\ \mu$ to $1\ \mu$);

(ii) particles of soot produced by cracking of the hydrocarbons in fuel-oil and gas flames ($0.01\ \mu < \phi < 0.1\ \mu$).

According to the methods of measurement employed (29)(30), the classification of the particles is made on the basis of the linear dimension, the specific surface or the volume. These three quantities being connected as a rule by very complex relationships in view of the shape of the particles, the possible existence of pores, the gloss on the surface, etc., it is not possible to define a powder completely by one method alone, except when the form of the particles is known and when this form is geometrically simple. In practice this amounts to saying that if possible a method of analysis must be chosen which classifies the particles directly according to the quantity which influences the phenomena we wish to study. It should be added that with the exception of microscopes, all the commercially available equipment carries out the classification by comparison with reference particles, the variety of which adds to the confusion. Moreover, the results are always presented as a function of equivalent diameters, defined as being those of spheres of the same surface or the same volume.

Finally, an important stage in the analysis is the extraction and preparation of the sample, the representative nature of which must be preserved.

Sieving

This is the simplest and most widely used method. Commercial equipment provides automatic vibration of a series of superimposed sieves, the mesh diameters of which decrease from high to low in geometrical ratio (usually $(2)^{\frac{1}{2}}$). Preliminary tests enable the optimum weight of powder for introduction into the top sieve to be determined as well as the sieving time required to ensure satisfactory reproducibility of the measurements (normally better than about 1 %). The particles are obviously classified according to their linear dimension. In this dry state this method can only be used validly for particles which are larger than 40 μ, because of the agglomerations which affect the results in the case of fine particles (sieving in the dispersive liquid phase makes it possible to come down to smaller sizes). Sieving is quite satisfactory for the granulometric control analysis of the pulverized coal used in our furnaces.

Microscopy

Direct observation of particles is possible by microscopy with white light above 1 μ, and below this by the electronmicroscope (down to 0·001 μ). The powder to be analysed is suspended in a dispersion liquid, one drop of which is then placed on the object slide. The evaluation is made photographically by comparing the image of each particle to a series of circles (or another geometrical shape) with areas which increase in accordance with a geometrical series of ratio 2, engraved on a transparent graticule, and classifying it in the circle of equivalent surface. There is in existence an electronic device for carrying out the evaluation automatically (31)(32). When this is done manually, granulometric analyses by microscopy are very long and tedious. On the other hand the method is certain (within operator error), and requires only a very small quantity of powder. It is mainly employed for the calibration and control of other methods of analysis. We use an electron microscope for analysing soots.

Elutriation-centrifuging

We use the Bahco apparatus, illustrated in Fig. 7.3: the particles are introduced into a current of air in the form of a flat convergent spiral and are separated into two fractions, one of larger particles which are thrown outward by centrifugal force, and the other of fine particles drawn to the centre by drag force. The dimension corresponding to the division into fractions is a function of the weight (volume × density) and the shape of the particles, and of the velocity of the current of air (magnitude and direction) at the point where the particles are injected. The equipment comprises a unit turning at constant speed (3000 rev/min) through which air is drawn by a turbine (T) having radial vanes (Fig. 7.3). The flow of air, and therefore its velocity, in the measurement chamber C is regulated by modifying the opening of the entry slit E by means of interchangeable shims introduced at F (the tangential velocity of the air remains close to that of the walls). The particles are injected into the separation chamber through a circular slit which is fed axially through the cone D. The fine particles are collected in A, and the large in B. The latter are retrieved, weighed and re-introduced into the apparatus after the opening E has been enlarged. In this way the granulometric analysis is achieved by means of 8 shims, between 3·5 and 63 μ for a solid of density 1. In the case of a body

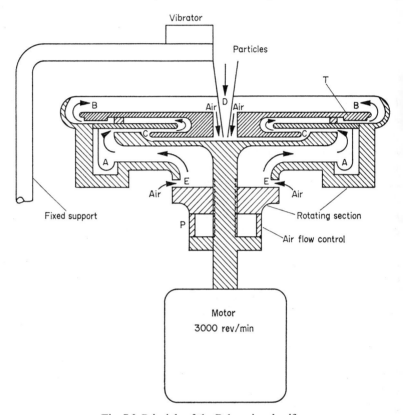

Fig. 7.3 Principle of the Bahco size classifier

whose density differs from 1 the dimensions of the division into fractions given by the maker, and defined for $d = 1$, are to be multiplied by the quantity $1/(d)^{\frac{1}{2}}$. The construction of the rotating unit ensures a regular and symmetrical flow of air and particles; the sources of error are mainly the variation in the rate of rotation and the characteristics of the air (temperature and pressure) in the separation chamber. It is generally accepted that the error in the fractionated diameter is normally less than 1%, and that the errors in weight are negligible when 5 g to 10 g of powder are introduced. The duration of a complete analysis is about 1 hour. The

Bahco apparatus serves as a complement to sieves for granulometric analysis of pulverized coal.

Coulter counter

This is a device which counts particles having volumes in excess of a given value; it can be used for particles with diameters of 1 μ to 200 μ.

Principle. Two electrodes immersed in a conducting saline solution (Fig. 7.4) in which the solid to be analysed is dispersed, are supplied with a stabilized electric current. One of these electrodes is insulated in a glass tube pierced at the bottom with a calibrated hole through which the solution is drawn. When a particle passes through the hole it modifies the resistance of the bath between the two electrodes and causes a pulse of potential difference of which the height e is, to a first approximation, proportional to the volume of the particle.

$$e = I\rho_0 \frac{(1 - \rho_0/\rho)}{S^2} V \times \left(1 + A_1(1 - \rho_0/\rho)\frac{s}{S} + A_2(1 - \rho_0/\rho)^2 \left(\frac{s}{S}\right)^2 + \ldots\right)$$

$$(7.1)$$

where: I = current between the two electrodes;

ρ_0 = resistivity of the electrolyte;

ρ = resistivity of the particle;

V = volume of the particle;

S = surface of the hole;

s = surface of the projection of the particle in the plain of the hole;

A_1 and A_2 = parameters depending on the shape of the particles.

The pulse is amplified and counted by an electronic device if its height exceeds a given threshold. The duration of counting is that of the passage of a known (or reproducible) volume of solution through the calibrated hole. In this way one can obtain the number of particles whose volume exceeds a fixed value and which are in suspension in a known volume of solution. Granulometric analysis is achieved by repeating the count for as many threshold values as required.

Equipment. This equipment comprises, basically,

(a) a glass tube pierced with a calibrated hole, the diameter of which is between 19 μ and 400 μ according to the requirements of the analysis. This tube is immersed in a glass container holding the solution for analysis;

(b) an agitator which keeps the suspension homogeneous. Its speed must be high enough to compensate for the differences in density between the

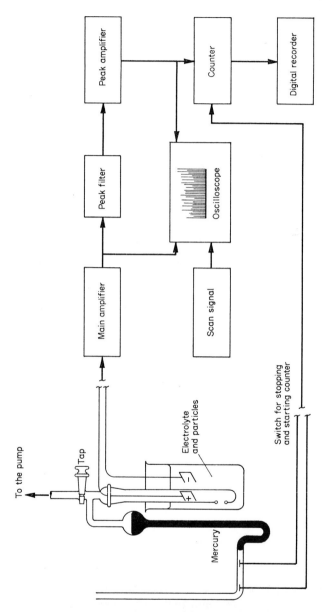

Fig. 7.4 Principle of the Coulter counter

solution and the particles, but not too high, so as to avoid the inertia forces of the large particles preventing their being drawn through the calibrated hole;

(*c*) a viewer for watching the flow in the calibrated hole;

(*d*) an electronic device for counting pulses which shows the result in numerical form. The threshold level above which the pulses are counted and the amplifier gain can be adjusted. A monitoring oscilloscope displays the peaks;

(*e*) a pneumatic device for ensuring the reproducibility of the volume of solution passing through the calibrated hole during the operation of the pulse counter. This operates as follows: with the cock R open, a pump draws in the solution and causes the mercury in the right-hand side of the manometer to rise. The cock is then closed and as the mercury falls it continues to draw in the solution. Two electrical contacts are arranged in the left-hand side of the manometer; the first controls the starting, and the second the stopping of the counter by the passage of the mercury; the volume of solution which has traversed the calibrated hole during this time is equal to the volume of the column of mercury between the two contacts.

The user selects the electrolyte, the tube diameter and the method of calibration as a function of the data of his particular problem.

(i) The electrolyte should have a density which is as near as possible to that of the particles, and an electrical conductivity which is suitable for bringing out the peaks quite clearly and ensures that the liquid is not heated by the Joule effect during its passage through the hole. In addition it must disperse fine particles: a dispersive liquid such as the Nodinet 22 can be added for this purpose. Batch (**33**) recommends a solution of 0·005 mole/1 of $No_4P_2O_7$ (sodium pyrophosphate) for the analysis of the solids from pulverized coal flames, in preference to a solution of NaCl (sodium chloride) which does not disperse fine particles satisfactorily. This author has not verified the chemical action or the dissolving of the particles in a solution of this kind.

(ii) The diameter of the hole is a function of the granulometry of the powder to be analysed. With a given diameter ϕ, the dimensions of the particles which can be measured fall between,

(*a*) a lower limit of about 0·02 ϕ due to the background noise;

(*b*) an upper limit which varies according to the shape of the particles, due to the error introduced by the correction term,

$$A_1 \left(1 - \frac{\rho_0}{\rho}\right) \frac{a}{A} + A_2 \left(1 - \frac{\rho_0}{\rho}\right)^2 \frac{a^2}{A^2}$$

of formula (7.2). This error is about 5% on a diameter of 0·40ϕ.

The limitations often result in the granulometry being too wide to be determined completely with one tube only. It is necessary then to separate the powder into two (or three) lots by filtering in the liquid phase, each lot being analysed with a different tube.

(iii) The calibration consists in relating the height h of the peak to the volume of the particle V or, more often, to the equivalent diameter d_a that the particle would have if it were spherical. According to equation 7.1, $h = KV = Kd_a^3$, (K is a calibration constant) so long as the measurements are made with sufficiently low values of s/S (negligible correction term). In addition, experience shows that the thermal conductivity of the particle has very little effect on the value of the constant K. Batch (**33**) explains this phenomenon, which is in contradiction to the basic equation, by the existence of a highly resistant absorbed layer of gas round the particle. Finally, the constant K is simply calculated from the height of the peaks given by particles of known diameter (measured with a microscope). There are certain powders available commercially which have a very restricted granulometry (pollen) and can be used directly, but greater accuracy is obtained by carrying out the calibration with the material it is desired to analyse. For this purpose it is necessary to prepare a powder of as restricted a granulometry as possible.

Once it is calibrated, the counter gives the number of particles having a volume greater than a given value, using certain corrections which take into account any possible coincidences (simultaneous passages through the calibrated hole). By repeating the measurement with a slightly lower threshold we can find by differences the number of particles having an equivalent diameter which falls between two sufficiently close limits for their total volume, and therefore their weight, to be calculated quite accurately. We can gradually establish a granulometric distribution curve which is generally in the form of a cumulative percentage by weight as a function of the equivalent diameter.

In practice, one is short of the weight of the residue of small particles which cannot be counted, a weight which has to be estimated if, as is usually the case, the concentration of solids in suspension in the electrolyte, and the volume drawn in are not known accurately. The best method is to choose the total volume of particles so that the resulting distribution corresponds to a particular law which is assumed to be valid in the region of small particles. When the distribution law for the powder is not known, the number of particles already counted can simply be extrapolated.

The duration of the operations for an analysis with about ten points is roughly half an hour, and the quantity of powder required is very small. The accuracy, under normal operating conditions, is better than 5% for the diameters. The result is that the error in the cumulative percentage by weight in the case of a pulverized coal analysis does not exceed 3 % relative (33)(34). The Coulter counter is, without any doubt, the most interesting modern device for granulometric analysis of particles sampled from the flames of pulverized coal.

Bibliography

1. LAIDLER, K. J. (a) *Chemical kinetics*. New York, etc., McGraw-Hill, 1950; (b) *The chemical kinetics of excited states*, Oxford, Clarendon Press, 1955.
2. KISSEL, R., LEVEQUE, M., RIVIERE, M., URBAIN, G. *Etude de l'évolution chimique de la combustion dans les flammes de diffusion par détermination des constituants des produits de combustion et interprétation des résultats*. I.F.R.F., Doc. nr. F 31/a/3, November, 1954.
3. LOISON, R. Les combustions hétérogènes: La combustion du charbon pulvérisé *Revue Générale de Thermique*, **58**, pp. 961–972, November, 1966.
4. BADZIOCH, S. (a) Correction for anisokinetic sampling of gas-borne dust particles; *Journal of the Institute of Fuel*, March, 1960; (b) Collection of gas-borne dust particles by means of an aspirated sampling nozzle. *British Journal of Applied Physics* **10**, pp. 26–32, January, 1959.
5. VITOLS. Theoretical limit of errors due to anisokinetic sampling of particulate matter. *Journal of the Air Pollution Control Association*, **16**, 2, February, 1966.
6. WALTER, E. Zur Problematik der Entnahmesonden und der Teilstromentnahme für die Staubgehaltsbestimmung in strömenden Gasen. *Staub*, **53**, pp. 880–889, 1957.
7. WHITELEY, A. B. and REED, L. E. The effect of probe shape on the accuracy of sampling flue gas for dust content. *Journal of the Institute of Fuel*, **32**, 222, pp. 316–320, July, 1959.
8. KISSEL, R. *Prélèvements des particules solides et des gaz dans les flammes*. I.F.R.F., Doc. nr. F 72/a/3, January, 1958.
9. KISSEL, R. *Appareils de mesures actuellement utilisés ou en cours de mise au point pour l'étude des flammes à IJmuiden*. I.F.R.F. Doc. nr. F 72/a/4, January, 1960.
10. CHEDAILLE, J. and BRAUD, Y. Equipements, méthodes et instruments nouveaux mis en service à la Station d'IJmuiden en 1964 et 1965. *6ème Journée d'Etudes sur les Flammes*, Paris, November, 1965.
11. BRAUD, Y. *Essais 0–15: Instruments de mesures utilisés pour l'étude des flammes incidentes vers la sole*. I.F.R.F., Doc. nr. F 72/a/10, July, 1966.
12. CHEDAILLE, J. and BRAUD, Y. *Essais préliminaires à C-12: Choix des flammes et mise au point des instruments*. I.F.R.F., Doc. nr. F 32/a/34, March, 1966.
13. BRAUD, Y. *Possibilité d'application de la spectrométrie de masse à l'étude des flammes*. I.F.R.F., Doc. nr. F 72/a/11, January, 1967.
14. BURTON, J. *Pratique de la mesure et du contrôle dans l'industrie, Tôme III*. Paris, Dunod, 1964.
15. VAN ZADELHOFF, G. A. *Appareil d'analyse de gaz Orsat*. I.F.R.F., Doc. nr. Tb-C 72/a/4, December, 1961.
16. TRANCHANT, J. *Manuel pratique de chromatographie en phase gazeuse*. Paris, Masson, 1964.
17. GREENE, S. A. and PUST, H. Use of silica gel and alumina in gas-adsorption chromatography. *Analytical Chemistry*, **24**, 7, July, 1957.
18. KYRYACOS, G. and BOORD, C. F. Separation of hydrogen, oxygen, nitrogen, methane and carbon monoxide by gas adsorption chromatography. *Analytical Chemistry*, **24**, 5, May, 1957.

174   *Measurements of Gas and Solids Concentration*

19. SCHOEDLER, G. Analyse automatique des gaz de hauts fourneaux par chromatographie en phase gazeuse. *Science et Technique,* **87,** 2, February, 1962.
20. WEBER, J. *Analyse des gaz issus d'une cuve de réduction.* Doc. Irsid, Dpt. Chimie Physique, July, 1964.
21. FARRE-RUIS, F. and GUIOCHON, G. Analyse rapide par chromatographie en phase gazeuse: séparation du mélange O_2, N_2, CH_4, CO. *J. Chromatographie,* **16,** pp. 382–390, 1964.
22. BRENNER, N. and CIEPLINSKI, E. *Gas chromatographic analysis of mixtures containing O_2, N_2 and CO_2.* Annals of the New York Academy of Sciences.
23. BUZON, J. and FOCCAIN, G. Analyses d'hydrocarbures légers par chromatographie gazeuse. *Revue de l'Institut Français du Pétrole,* June, 1961.
24. TAYLOR, C. W. and DUNLOP, A. S. The analysis of light hydrocarbons by gas-liquid chromatography. In: Coates, Noebels and Fagerson: Gas Chromatography, a Symposium held under the auspices of the Instrument Society of America.
25. DE WET and PRETORIUS. Some factors influencing the efficiency of gas liquid partition chromatography columns. *Analytical Chemistry,* **30,** 3, pp. 325–332, 1958.
26. ROBERTS, A. L. and WARD, C. P. *Application of modern analytical techniques in the gas industry. Part II: The complete analysis of fuel gases by gas chromatography.* London, Gas Council Research Communication G. C. 114. 30th Autumn Research Meeting of the Institute of Gas Engineers, November, 1964.
27. NACHMANSOHN, B. and SOULÉ, C. Récents progrès en chromatographie: la détection h.f. à effluves. *Mesures,* 288, pp. 601–610, May, 1961.
28. BEVAN, S. C. and THORBURN, S. *Absolute mass integral detector for gas chromatography.* Chemistry in Britain, pp. 206–208, 1965.
29. LEENAERTS, R. Les techniques d'analyse granulométrique—Principes et appareillage. *Revue Universelle des Mines,* 8, pp. 197–207, August, 1966.
30. CLYDE ORR JR. and DALLAVALLE, J. M. *Fine particle measurement.* New York, McMillan, 1959.
31. FISHER, C. *The metals research image analysing computer.* (Particle Size Analysis Conference, Loughborough University of Technology, September, 1966.) London, Society for Analytical Chemistry.
32. COLLINS, G. F. *Some tests of the Mullard, type L 188, automatic particle analyser.* (Particle Size Analysis Conference, Loughborough University of Technology, September, 1966), London, Society for Analytical Chemistry.
33. BATCH, B. A. The application of an electronic particle counter to size analysis of pulverized coal and fly-ash. *Journal of the Institute of Fuel,* **37,** pp. 455, October, 1964.
34. ALLEN. *A critical evaluation of the Coulter-counter.* (Particle Size Analysis Conference, Loughborough University of Technology, September, 1966), London, Society for Analytical Chemistry.
35. LE DINH, P. H. L'épuration industrielle de l'air. *Revue Générale de Thermique,* 51, pp. 1081–1107, November, 1966.

Part 4

Velocity Measurement

Introduction

The accurate measurement of velocity magnitude and direction plays an important part in the study of combustion and heat transfer in industrial furnaces, since it allows:

(i)   the establishment of mass balances for the determination of combustion rates and re-circulation flow rates;
(ii) the determination of the position and the flow patterns of direct jets (primary and secondary) and re-circulated products;
(iii) isokinetic sampling of gas and solid particles; and
(iv) calculation of the coefficients of heat exchange by convection.

The method used for the measurement of velocity is based upon the Bernoulli equation for an incompressible flow. For jets without swirl the Prandtl tube is generally used for the measurement of velocity magnitude, the velocity direction being parallel to the furnace axes to within $\pm 10°$. When the direction as well as the magnitude of the velocity is to be determined, a tube with a hemispherical head containing 5 holes is used (1)(2). Even though these methods may be considered as classical, the severe conditions found in industrial furnaces make their application difficult. The principal problems encountered are:

(a) mechanical resistance of probes to very high temperatures;
(b) the measurement of very small differential pressures of the order of 1 mm water gauge;
(c) the blockage of orifices in instruments by particles carried in suspension;
(d) the effect of these particles and the level of turbulence, generally very high, on the measurement of pressure.

Because of these difficulties the classical instruments have had to be modified and in certain cases have had to be redesigned. In this Part we shall discuss the fundamental aspects of the problem of velocity measurement, possible sources of error and correction formulae.

Chapter 8

The Pitot Method

8.1 Principle

The Bernoulli equation is given by

$$p + \rho g Z + \frac{\rho V^2}{2} = \text{constant}$$

This equation expresses the conservation of energy along a streamline in a perfect fluid (non-viscous and incompressible), and is used in particular for the fluid element impinging on the stagnation point A on an obstacle placed in a uniform flow (Fig. 8.1 (a)).

$$p_A + \rho g Z + 0 = p_S + \rho g Z + \frac{\rho V_0^2}{2} \qquad (8.1)$$

where p_A is the pressure at the stagnation point, the total pressure;

p_S is the pressure in the flow at the same height, the static pressure;

Z is the height of the measuring point and

$\rho g Z$ represents the potential energy;

V_0 is the main stream flow velocity;

$\dfrac{\rho V_0^2}{2}$ is called the dynamic pressure and

$\dfrac{V_0^2}{2g}$ is referred to as the velocity head.

From equation 8.1 the velocity of flow from V_0 can be determined from the measurement of the pressure differential $p_A - p_S$:

$$V_0 = \sqrt{\frac{2(p_A - p_S)}{\rho}}$$

The Pitot tube is based upon this principle.

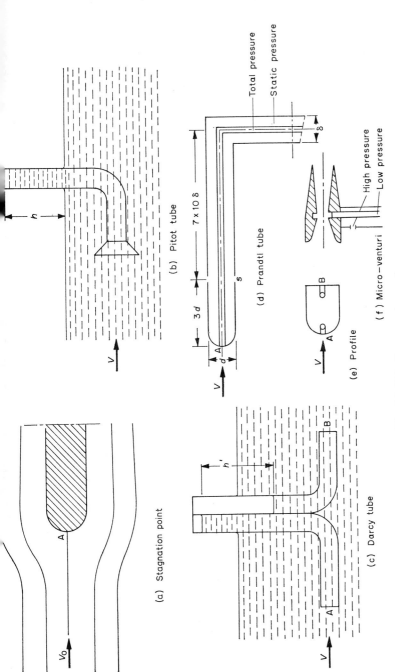

(a) Stagnation point

(b) Pitot tube

(c) Darcy tube

(d) Prandtl tube

(e) Profile

(f) Micro-venturi

Fig. 8.1 Velocity measuring probes

8.2 Probes

The Pitot tube was initially constructed for measurements of velocity in water (Fig. 8.1 (b)). It was a simple tube bent at right angles facing the flow giving an over-pressure h above the free surface of the water such that

$$h = \rho \frac{V^2}{2} \quad \text{or} \quad V = \sqrt{\frac{2h}{\rho}}$$

Darcy improved this tube (Fig. 8.1 (c)), by adding a second tube bent at right angles and facing downstream which measures the pressure, $p_B = k(\rho V^2/2)$, so that the differential pressure between the two tubes is given by

$$p_A - p_B = h' = (1 + k)\frac{\rho V^2}{2}$$

which gives
$$V = \sqrt{\frac{2h'}{(1 + k)\rho}}$$

Prandtl then proposed a tube (Fig. 8.1 (d)), in which the constant k is not needed, by a judicious placing of the second pressure tapping that measures the static pressure so that

$$p_A - p_S = \frac{\rho V^2}{2} \quad \text{or} \quad V = \sqrt{\frac{2(p_A - p_S)}{\rho}}$$

The tubes that are generally used for the measurement of velocity are Prandtl tubes or forms of this tube except for certain particular measurements which require more compact tubes. A streamlined tube is used (Fig. 8.1 (e)) (3), consisting of an upstream pressure tapping A and a downstream pressure tapping B such that

$$p_A - p_B = (1 + k)\frac{\rho V^2}{2}$$

The constant k has therefore to be measured by a previous calibration. For low density fluids (hot gases) these probes have the disadvantage that they only give small pressure differentials. There is the possibility of increasing the sensitivity of the instrument by using a micro-venturi (Fig. 8.1 (f)), for which the multiplication coefficient, which depends on a number of factors, has to be determined by calibration (4).

8.3 Positioning of pressure tappings

Flow around an obstacle—influence of Reynolds number

In practice, the viscosity of the gas is not zero (Table 8.1) and increases rapidly with temperature according to the formula of Sutherland.
  Dynamic viscosity

$$\mu = \frac{kT^{3/2}}{T + C}$$

with k and C constants dependent upon the nature of the gas.

  Kinematic viscosity $v = \dfrac{\mu}{\rho}$

As a result a boundary layer is formed over the surface of the obstacle. Within this boundary layer, energy is dissipated by friction and when the adverse pressure gradients become sufficiently large separation of the flow occurs with the formation of a turbulent zone in the wake of the obstacle where the Bernoulli theorem is no longer applicable (5).

  For a cylinder with its axis perpendicular to the flow, several turbulent flow regimes are set up (Fig. 8.2) depending upon the value of the Reynolds

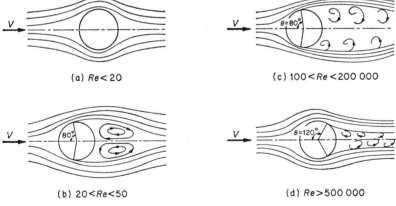

(a) $Re < 20$

(c) $100 < Re < 200\,000$

(b) $20 < Re < 50$

(d) $Re > 500\,000$

Fig. 8.2 Flow around a cylinder with access perpendicular to main flow direction

number, the similarity criterion ($Re = Vd/v = \rho\,Vd/\mu$) which is the ratio of the inertia forces to the viscous forces.

(*a*) For very small Reynolds numbers, (*Re*) less than 20, the influence of viscosity is dominant on the flow and Stokes has shown that the streamlines are the same as those for a potential flow of a perfect fluid (Fig. 8.2 (*a*)).

7

(*b*) When the Reynolds number increases up to 20, there is a thickening of the boundary layer followed by the formation of a wake, of the same diameter as the obstacle, enclosing two stable vortices. Separation occurs slightly upstream ($\theta = 80°$) (Fig. 8.2 (b)). As the Reynolds number is increased the centres of the vortices move away from the obstacle and the volume of the wake increases.

(*c*) When the Reynolds number reaches a critical value (between 50 and 100), the vortices break away from the body and give rise to a system of two symmetrical alternating vortices known as the Von Karmann vortex street. Separation still takes place upstream of $\theta = 80°$ (Fig. 8.2 (c)).

(*d*) When the Reynolds number passes its second critical value between 200 000 and 500 000, the point of separation moves abruptly downstream to $\theta = 120°$ (Fig. 8.2 (d)). The wake also becomes more narrow. This is because, according to Prandtl, the laminar boundary layer becomes turbulent upstream of the separation point and therefore the dissipation of energy is weaker. It is also possible to use this regime for Reynolds numbers less than 200 000 by artificially creating turbulence on the upstream face of the cylinder.

These different flow regimes are also found for flow over spheres, and other obstacles with rounded contours. It should be appreciated therefore that a velocity measuring tube calibrated for one flow regime cannot be used without correction in another flow regime. In an industrial furnace for a tube 10 mm in diameter, the conditions are in general those of regime (c) ($500 < (Re) < 20\,000$) but when the temperature is very high (1500°C) and the velocity is low (1 m/s) flow may be in region (b).

Pressure distribution around an obstacle

For the determination of velocity by the Pitot method, one is concerned essentially with the pressure distribution over the surface of the obstacle. Pressure distributions have been measured for various shapes, in particular for a cylinder with its axis perpendicular to the flow, for a sphere and for an aerofoil-shaped body (Fig. 8.3) (5)(6). It can be seen that for Reynolds numbers that are sufficiently large, the dynamic pressure is measured at the forward stagnation point and the pressure distribution on the upstream face of the obstacle, up to $\theta = \pm 35°$, remains close to or coincides with that of a perfect fluid (theoretical potential flow satisfying the Bernoulli theorem). As the separation point is approached, and particularly over the downstream face of the obstacle, the flow conditions differ considerably from that of potential flow according to the different flow regimes defined above. It should be noted that for short obstacles, the pressure on the downstream face remains practically constant over a large area so

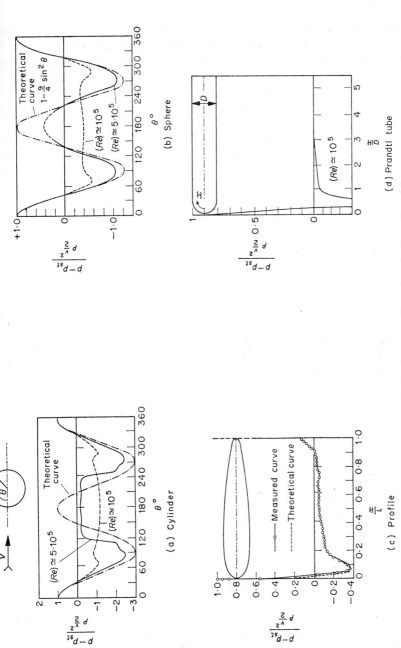

Fig. 8.3 Pressure distribution around an obstacle (from Prandtl)

that a pressure tapping placed in the centre of this area gives a reading which is largely insensitive to probe orientation (streamlined tube of the Foundation). There is also one point on the surface where the pressure is equal to the static pressure. For a sphere and a cylinder, this point is situated in a zone of large pressure gradient ($\theta = 35°$ approximately). For an elongated body, however, this point is found towards the back of the body in a region of small pressure gradient. In this latter case the static pressure tapping can be fixed without risk of a significant error as a result of inaccurate positioning of the pressure tapping with respect to the tube axis, or for the tube with respect to the velocity direction. The Prandtl tube is based upon this principle.

Standardization of velocity measuring tubes

We have seen above that for the flow regime (c) ($100 < (Re) < 200\,000$) the total pressure (dynamic + static) is measured at the forward stagnation point of the tube. It now remains to determine the diameter d of this

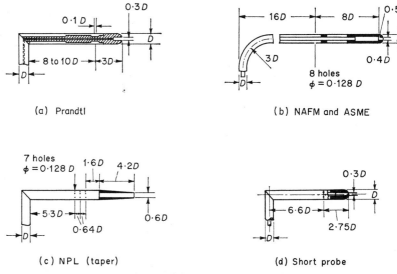

(a) Prandtl (b) NAFM and ASME

(c) NPL (taper) (d) Short probe

Fig. 8.4 Standardization of velocity measuring tubes

pressure tapping in relation to D the diameter of the tube. Experiments have shown (7) that for a tube with a hemispherical head there is no significant error when the ratio d/D is between 0·2 and 0·74. For values of d/D less than 0·2 the position of the pressure tapping on the tube axis and the orientation of the tube along the velocity direction must be carefully

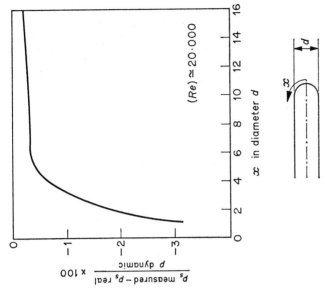

(b) Pressure curve along the tube (from Ower and Johansen [**32**])

(a) Influence of beam and orientation of hole (from Hubbard [**10**])

Fig. 8.5 Position for static pressure holes

adjusted in order to avoid measuring a pressure lower than the total pressure. For values of d/D larger than 0·74, the disturbance of the flow by the tube becomes significant so that it becomes more difficult to measure the static pressure simultaneously with the total pressure. The ratio d/D has been fixed for different types of tube heads (8) according to the particular measuring conditions (Fig. 8.4). The position of the static pressure tappings on the tube depends upon the form of the tube head and the position of the support for the tube head. A number of experiments such as those shown on Fig. 8.5 (7)(8) have shown that static pressure should be measured by a number of holes uniformly distributed around the circumference of the tube and placed at a distance sufficiently far from the support of the probe head. The pressure tapping has also to be sufficiently far from the head of the tube in order to effectively measure the static pressure, *i.e.* at a position where the streamlines are re-attached to the tube wall and where the velocity is equal to that of the measured velocity. For example, for a tube with a hemispherical head of diameter D, the minimum distance is $x = 2·5D$. Standards have been fixed for the construction of tubes for which the constant $k = \Delta p/(\rho V^2/2)$ remains equal to 1, Fig. 8.4.

If any modifications are made to the tube it is necessary for the tube to be calibrated. This is the case, for example, for the streamlined tube used at IJmuiden (3), or the sword blade tube (9) which has been constructed for velocity measurements in regions where the Prandtl tube is too cumbersome. The second pressure tapping is placed in the wake of the tube which is aligned perpendicular to the flow direction.

Streamlined tube: $$1·60 > k = \frac{\Delta p}{\rho V^2/2} > 1·25$$

Sword-blade tube: $$1·90 > k = \frac{\Delta p}{\rho V^2/2} > 1·75$$

Pressure tappings

Static pressure tappings. If a cavity is made in the wall of a duct through which fluid is flowing, as shown in Fig. 8.6 (a), the pressure set up inside this cavity is the static pressure of the surroundings, provided that no vortex is formed inside the cavity. The dead fluid filling the cavity may be considered as an element of the wall and it is sufficient to connect a manometer to the cavity in order to measure the static pressure. If a burr is formed at the edge of the hole, or if the wall in the proximity of the hole is not perfectly smooth, the local flow conditions will be altered and consequently the measured static pressure. It is important therefore that the edges of the hole should be perfectly smooth (Prandtl considered that a small rounding of the hole was permissible) and the surface of the wall in the vicinity of the hole should have no projections perpendicular to the flow direction. In practice it is preferable to construct separately the pressure tapping in a carefully finished plate which is then fitted to the wall rather than making a hole directly into the wall (5). For measurement of

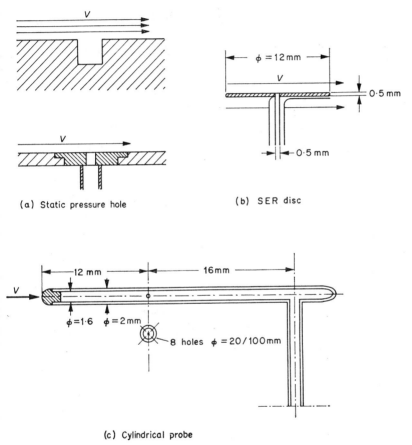

(a) Static pressure hole

(b) SER disc

(c) Cylindrical probe

Fig. 8.6 Static pressure holes

the static pressure at a point in a flow, a disc, based upon the principles described above, can be used. This is equivalent to an element of the wall and should be as thin as possible. The disc must be aligned in the plane of the flow direction and in order to make this adjustment a yaw meter is sometimes fitted near the edge of the disc (Fig. 8.6 (b)). If the disc is not properly aligned then an error results according to whether the disc is facing upstream or downstream the flow. The cylindrical static pressure probe (Fig. 8.6 (c)), is more convenient for use and is insensitive to errors of orientation of plus or minus 5°, but it is less precise in turbulent flows **(10)**. When this probe is at an angle θ to the flow direction the pressure reading is too low and for a tube with a hemispherical head **(11)** is given by

$$\frac{p_{Sm} - p_S}{\rho V^2/2} = K_s \sin^2 \theta$$

where K_s is positive and p_{Sm} is the measured static pressure.

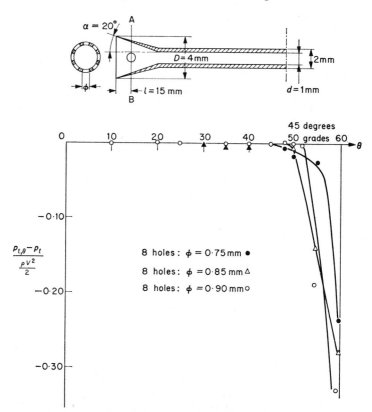

Fig. 8.7 Total pressure probe insensitive to velocity orientation

Total pressure. We have already seen that the total pressure (dynamic + static) is that measured at the upstream stagnation point of an obstacle. It is therefore, sufficient simply to make a hole and connect this to a manometer. Since the velocity at this point is zero, the condition of the surface in the vicinity of the pressure tapping does not have to be particularly smooth. We have also seen that the influence of the ratio, d/D, has to be fixed for each type of instrument. For large values of this ratio the probe is less influenced by the direction of the flow and probes such as those shown in Fig. 8.7 can be used for making measurements where the angle θ is large. (The error is negligibly small for θ less than 45°.) Also a probe with a very small diameter (2 mm) gives a measurement which is virtually at a point and this is particularly useful in turbulent flows.

8.4 Factors influencing the measurement

Reynolds number

For low Reynolds numbers viscous forces are no longer negligible in comparison to inertia forces and the Bernoulli theorem is no longer applicable.

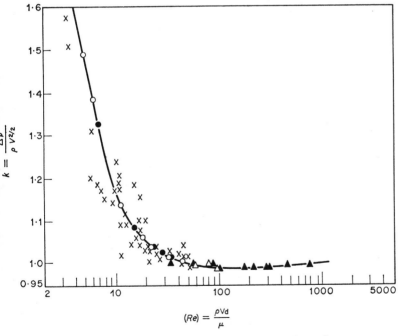

Fig. 8.8 Calibration of Prandtl tube for low Reynolds numbers

This is the case for flows of type (a) and (b). The method of Pitot remains valid provided that the measured pressure is corrected by dividing by a factor $k = \Delta p/\rho(V^2/2)$ which has been calculated directly by a number of authors

(9), for simple geometric forms and within well defined Reynolds number ranges (Table 8.2). The factor increases very rapidly when the Reynolds number is less than 50. Even though the theoretical formulae are in close agreement with experimental results it is recommended that a calibration be made for each tube, if it is desired to achieve a high precision for velocities at low Reynolds numbers. Figure 8.8 shows some curves which have been obtained by this method (7)(12)(13). In industrial furnaces where the temperature is high correction needs to be made to measurements of very low velocities (less than 1 metre per second) which would otherwise give a reading which was too high.

Compressibility of gas

The Bernoulli theorem cannot be applied when the conditions in the gas can no longer be considered as incompressible. For measurements at high velocities the readings must be corrected (7). For diffusion flames in industrial furnaces, velocities are rarely higher than 80 metres per second and therefore this problem will not be considered here.

Orientation of the probe

As we have seen in the section on pressure tappings on page 186 the velocity measuring tubes have directional properties. The case which has been most investigated is that where the total pressure tapping is placed in a spherical head or in the hemispherical head of a Prandtl tube.

We have seen that when considering the pressure distribution around an obstacle (page 182), that provided the angle θ is not too great, then the velocity field and consequently the pressure field is practically the same as that of a potential flow, and that the error is of the form

$$p_{t,\theta} - p_t = -K \frac{\rho}{2} V^2 \sin^2 \theta$$

where K is equal to 9/4 for a sphere and 1·9 approximately for a hemispherical head and θ is the angle between the velocity direction and the probe. It has also been shown (11) that the static pressure around a cylinder is of the form

$$p_{s,\theta} - p_s = -K_s \frac{\rho}{2} V^2 \sin^2 \theta$$

where K_s depends upon the number and the arrangement of the static pressure holes on the tube. For the probe shown in Fig. 8.6 (c) this coefficient has the value 0·16. (Fig. 8.9 shows the errors that can arise when pressures are measured for the Prandtl tube which is at an angle to the flow.)

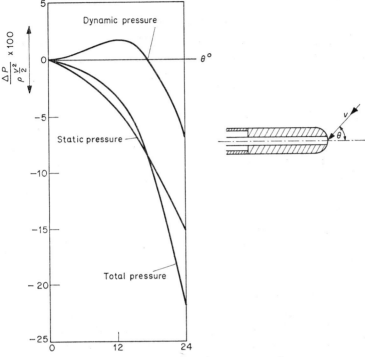

Fig. 8.9 Influence of probe orientation on measured pressures

Turbulence

In industrial furnaces where there is a mixing of gases at different temperatures by turbulent diffusion the fluctuations of velocity in magnitude and direction, particularly near the edges of the jets, are of great importance. Measurements are aimed in general at determining the time mean average $\bar{V}$ of the velocity which is given by the expression

$$\bar{p}_t = \bar{p}_s + \rho \frac{\bar{V}^2}{2} + \frac{\rho}{2} (\overline{u'^2} + \overline{v'^2} + \overline{w'^2}) \qquad (8.2)$$

where $\bar{p}_t$ and $\bar{p}_s$ are the mean total and static pressures respectively and u', v', w' are the instantaneous values of the velocity fluctuations in the average direction of the velocity and in two axes perpendicular to this direction. A comprehensive study of this subject has been made by Barat (11) and we discuss here the principal results.

Static pressure probe. We shall consider the particular case of the probe shown in Fig. 8.6 (c). The turbulence is supposed to be homogeneous and isotropic ($\overline{u'^2} = \overline{v'^2} = \overline{w'^2}$) so we may write

$$\bar{p}_{sm} = \bar{p}_s + K\rho\overline{v'^2}$$

where $\bar{p}_{sm}$ is the measured static pressure. The value of the coefficient K depends on the dimension λ of the small eddies, the transverse correlation length L characterizing the turbulence and the diameter of the probe D. Three cases are considered:

(1) D greater than L: this is the case studied in detail by Goldstein **(14)** where there is no correlation between two diametrically opposed orifices for measurement of static pressure. In this case $0 < K < \frac{1}{2}$.
(2) $D < \lambda$: this is the opposite case to the previous one and the probe behaves as though it was inclined to the velocity; K is less than zero.
(3) $\lambda < D < L$: the sign of K depends on the energy distribution between the eddies.

The author has verified the validity of this classification experimentally by introducing probes of different diameter into a flow with a constant turbulence.

Total pressure probe. According to the equation 8.2 the total measured pressure is too high by $\rho(\overline{u'^2} + \overline{v'^2} + \overline{w'^2})/2$ as compared to the mean velocity $\bar{V}$. The Prandtl tube does not give a reading as high as this because of the directional properties of the probe and this makes the problem considerably more complicated. (Positioning of pressure tappings, page 181.)

Correction of measured pressures. We shall again consider the case of a Prandtl tube with a hemispherical head for which the error in total pressure and static pressure, due to an orientation of the probe to the flow in a uniform flow field, can be written as

$$\frac{p_\theta - p}{\rho V^2/2} = -K' \sin^2 \theta$$

where θ is the angle made by the axis of the probe with the velocity direction.

For an instrument placed along the mean flow direction, the instantaneous turbulent velocity has an angle $\theta(t)$:

$$\sin \theta(t) = \frac{\sqrt{v'^2 + w'^2}}{\sqrt{(\bar{V} + u')^2 + v'^2 + w'^2}}$$

i.e. the error in the instantaneous pressures is given by an equation of the form

$$\frac{p_0(t) - p_\theta}{\rho \bar{V}^2/2} = -2K' \frac{\bar{v}'^2}{\bar{V}^2}$$

The turbulence is assumed to be homogeneous and isotropic ($\overline{u'^2} = \overline{v'^2} = \overline{w'^2}$). The coefficient K' depends upon the type of pressure hole (total or static), the shape of the probe and the intensity of turbulence, according to the classification given in the paragraph on static pressure probes. This coefficient has to be measured experimentally and it is not possible to make the corrections unless the intensity of turbulence is known. As an example the author considers a jet with an initial diameter $2a$ at a distance $x = 40a$ from the ejection orifice. For a Prandtl tube 4 mm in diameter the percentage error is 5% low on the total pressure and 3·5% on the static pressure. At the edges of the jet these errors are more than doubled for an intensity of turbulence of 25%.

Correction to the velocity. The measured dynamic pressure $\bar{p}_{dm}$ is the difference between the measured total pressure p_{tm} and the measured static pressure $\bar{p}_{sm}$

$$\bar{p}_{dm} = \bar{p}_{tm} - \bar{p}_{sm}$$

For a Prandtl tube the gradient in static pressure is considered negligible so that the static pressure is also considered to be measured at the forward stagnation point. The author shows that when the tube is small in comparison to the diameter of the eddies the value of the constants (K') for the tube are approximately $K'_t = 1\cdot85$ and $K'_s = 0\cdot6$. The correction formula is then:

$$\bar{V} = \bar{V}_m \left(1 - 0\cdot25 \frac{\overline{u'^2}}{\bar{V}^2}\right)$$

If the total pressure measurement is not sensitive to the direction of the velocity when the intensity of turbulence is less than 25%, $K'_t = 0$ and the correction formula becomes

$$\bar{V} = \bar{V}_m \left(1 - 2\cdot1 \frac{\overline{u'^2}}{\bar{V}^2}\right)$$

Conclusion. Turbulence can be the cause of significant errors in the measurement of velocity. Corrections for these errors are difficult to make in practice since the intensity of turbulence is generally not known and is difficult to measure, and the constants K' cannot be easily determined.

Density of gas

In order to calculate the velocity from the pressure measurements the density of the fluid must be known, *i.e.* its composition, temperature and pressure. It is in general not possible to measure all these quantities at the same time and the measurements of the physical quantities at different times can only be used if the flow remains steady. It is possible to measure the static pressure at the same time as the pressure differential. The fluctuations of temperature and composition can be very significant in the initial region of a diffusion flame where combustion and mixing of the jets with the recirculation is taking place. These fluctuations are associated with the velocity and pressure fluctuations whose amplitude at times is greater than the mean value. The density of the gaseous mixture is given by the formula:

$$\rho = \rho_{01} \times \frac{273}{273 + t} \times \frac{p_a}{760}$$

where ρ_{01} is the density of the gas under normal conditions, p_a the atmospheric pressure and t the temperature of the gas in °C.

Solid particles in suspension

In industrial furnaces the flames are often laden with particles. In gas and oil flames these particles are very small (mean diameter less than 1 μ), and their concentration is usually small, of the order of a few milligrammes per litre of gas under normal conditions. In pulverized coal flames the size of particles lies between 1 μ and 200 μ and concentrations can be several grammes per litre of gas under normal conditions. These particles affect the measurements by direct deposition on the probe, blocking off the holes and burning. Account also has to be taken of changes in the density of the fluid due to the presence of the particles.

Deposition of particles on the probe. When measurements have been made with a Prandtl tube in a pulverized coal jet, large-scale deposition has been found on the head of the tube and under these conditions the measurement, if it can be taken at all, is incorrect. The pressure holes, particularly the total pressure hole, block progressively, causing a damping of the fluctuations and causing a drift in the reading. The tubes have to be cleaned by blowing through with compressed air. This operation is carried out by using a system of switches on the compressed air line by means of electromagnetic switches (Fig. 8.10). Care must be taken to protect the pressure transducer from high pressures during the blowing and after blowing: time

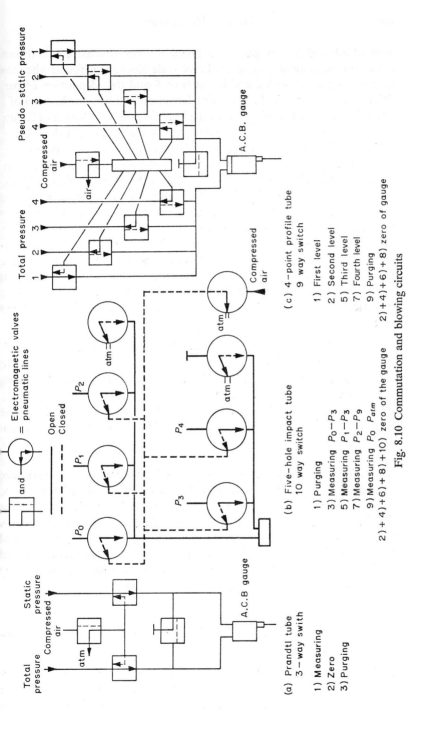

Fig. 8.10 Commutation and blowing circuits

= Electromagnetic valves
= pneumatic lines

Open
Closed

Pseudo-static pressure

1 2 3 4

Compressed air

air

A.C.B. gauge

Total pressure

1 2 3 4

Compressed air

(c) 4-point profile tube
9 way switch

1) First level
2) Second level
5) Third level
7) Fourth level
9) Purging
2)+4)+6)+8) zero of gauge

(b) Five-hole impact tube
10 way switch

P_0 P_1 P_2 atm

P_3 P_4 atm

Compressed air

1) Purging
3) Measuring $P_0 - P_3$
5) Measuring $P_1 - P_3$
7) Measuring $P_2 - P_9$
9) Measuring P_0 P_{atm}
2)+4)+6)+8)+10) zero of the gauge

(a) Prandtl tube
3-way switch

Total pressure Static pressure

Compressed air

atm

A.C.B gauge

1) Measuring
2) Zero
3) Purging

must be allowed for the gas in the pressure tubes to achieve equilibrium before the measurement is taken. Large-scale blockage is found when the tube is completely cooled and it has been found that deposition of particles is reduced, particularly the droplets of oil and tar, on the head of a Prandtl tube by not cooling part of the head (in practice a length of 25 mm or 35 mm for a tube of 19 mm in diameter of stainless steel 18Ni–8Cr).

The head of the tube is then heated by radiation to a temperature of approximately 900°C so that particles which deposit on the head are burnt immediately. A disadvantage of this system which is less subject to blockage, is on the one hand that the particles burn inside the pressure holes as well as outside, thereby resulting in a false measurement, and on the other hand the combustion of particles on the probe can change the local flow and temperature conditions and thereby affect the static pressure measurement. By this means the blockage has been practically eliminated in the pulverized coal jet and in the oil jets in the region between the burner exit and the flame front. We have also found that the principal factors causing increased blockage, in addition to an increase in the density of particles are:

(i) the velocity: as the velocity is increased the rate of blockage is increased;
(ii) the type of coal: coal containing high volatile matter was found to block very much more rapidly. This may be due to the particles becoming more sticky as the tar is released;
(iii) the dimensions of the tube and its form: smaller tubes have less tendency to block and hemispherical heads are not as good as conical heads.

Correction for the density of the fluid. (1) Static pressure: The static pressure holes are placed at a point where the fluid velocity is that which has to be measured, *i.e.* at a place where the flow is undisturbed. Under these conditions the particles move at the same velocity of the flow and follow the streamlines. Each particle, provided it is not too large (gravitational forces small in comparison to the aerodynamic forces), behaves as a fluid particle and the specific weight which has to be taken is the average specific weight, *i.e.* the total weight per unit volume;

$$\rho = \rho_{gas} + C$$

where C is the concentration of particles per litre of gas under the measuring conditions (**15**).

(2) Total pressure: The effect of the particles on the total pressure measurement is very much more complicated than that on the static pressure measurements since the total pressure measurement is dependent upon

he manner in which the particles are decelerated as they arrive at the stagnation point. Factors influencing this measurement are dependent upon the characteristics of the particles (dimensions, form, density, velocity and concentration), on the gas (density, viscosity and velocity) and of the probe (dimension, form, diameter of the pressure hole, etc.). It is generally accepted that the gas and particles are moving at the same velocity upstream of the tube. It is in general not possible to measure the velocity of the particles alone. Two extreme cases can arise in which the particles are sufficiently large so that they move independently of the streamlines, or alternatively they are sufficiently small that they behave as particles of fluid (16).

(a) The case of large particles is itself an intermediary stage between two extreme cases. The particles will lose their kinetic energy by collision or friction against the tube. The fluid will not recover this energy and the density to be taken in this case is that of the fluid itself. When the particles are decelerated by the fluid, the change in kinetic energy of the particles results in an increase in a total pressure as given by equation

$$p_t = \rho_G \, V^2/2 + p_s + CV^2$$

i.e.

$$p_t - p_s = \left(\frac{\rho_G + 2C}{2}\right) V^2$$

The mixture is thus considered to have a density $\rho = \rho_G + 2C$.

(b) When the particles are sufficiently small that they follow the streamlines and behave like elements of fluid there is no exchange of energy between the particles and the fluid. Kinetic energy is transformed to potential energy according to the Bernoulli theorem and the density to be taken is the average density, *i.e.* $\rho = \rho_G + C$.

(c) In practice for the very fine particles of soot in gas and oil flames, case (b) applies, and corrections are often negligible (less than 0·1% when concentrations are low of the order of a few milligrammes per litre). On the other hand, for pulverized coal flames in which particle sizes vary over a wide range, it is not possible to have a single correction formula $\rho = \rho_G + kC$ which is valid for all the points in the furnace and for all the instruments. At IJmuiden the value of $k = 1$ is chosen for the calculation of ρ.

Velocity measuring instruments in jets laden with particles. Since the Prandtl tube can be easily blocked special probes have been designed for measurements in fluids laden with particles.

8

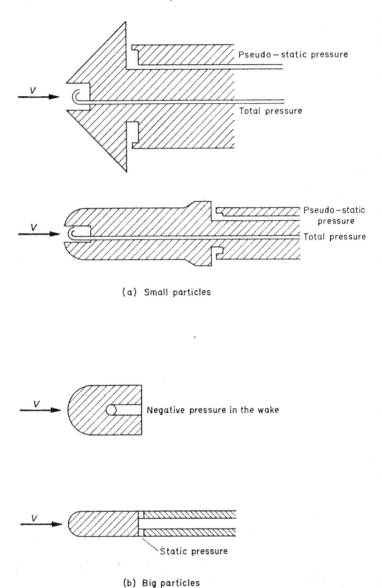

(a) Small particles

(b) Big particles

Fig. 8.11 Probes for measurements in dust laden gases

(*a*) Small particles and low concentrations:
Blockage of the total pressure hole can be avoided by the method shown
in Fig. 8.11. The small tube, bent so as to face downstream, is fitted into
a large cavity at the stagnation point of the probe head. The probe static

pressure is not measured, but a pseudo static pressure is measured by a pressure tapping protected from the particles by a flange. These probes have to be calibrated in a wind tunnel.

b) Large particles and high concentrations:

This is the case for measurements of velocity in cold jets before the flame front where the blockage of the total pressure hole takes place in a few seconds. The method adopted at IJmuiden is to measure the static pressure with the cylindrical tube, and subsequently to measure the depression in the wake of a tube. This tube also has to be calibrated.

Tests carried out in the region near the exits to the burner in the pulverized coal furnace showed that the concentration which should be elected when calculating the velocity should be the mean concentration:

$$ = \rho_G + C.$$

Temperature of the tube

Temperatures in industrial furnaces can be as high as 2000°C. Velocity measurements can be made with water-cooled probes, but as we have seen this can result in condensation of water vapour and droplets, as well as the deposition of particles.

Alternatively, probes that are not water cooled have been made with refractory materials (alumina) or stainless steel, in which deposits are burnt on the probe head. At IJmuiden, a combination is used in which a stainless steel head is left uncooled over a distance of 25 mm or 35 mm. In all cases there will be a difference in temperature between the probe and the gas. The temperature of non-cooled probes depends more on the intensity of radiation than upon the temperature of the gas. As a result of this temperature difference there is an exchange of heat by convection between the gas and the probe, and this causes a perturbation in the boundary layer, and therefore on the flow. Comparisons which have been made with cooled and non-cooled Prandtl tubes have not shown any significant differences (2).

Velocity gradients. It has been assumed up till now that the probe is placed in a uniform flow field. If this is not the case then the Bernoulli theorem is still applicable to the element of fluid impinging on the stagnation point. If a high degree of accuracy is required in the measurements, it is necessary to take into account that this point is not the geometrical centre of the head of the tube, but is displaced towards the position of maximum velocity (7). Errors can arise if the diameter D of the tube is too large and the ratio l/D is too small. Special flat tubes have also been designed for measurement of velocity in boundary layers (7).

Conclusion

The factors influencing the measurement of velocity are numerous and
can result in sources of error. There are, in the first instance, errors due
to the different methods of measurement, bad orientation of probes,
inaccuracies in the constants chosen for the probes and the pressure
transducers, non-linearity in the output of the transducers, errors in the
reading of recordings, etc. The resultant of all these errors, which are
evidently dependent upon the care with which the measurements are taken,
should not normally be more than a few per cent. Other errors can be
considered as being due to difficulties in measuring conditions, *i.e.* the
turbulence of the flow, particles in suspension, the importance of molecular viscosity of the fluid at low velocities and high temperatures, the
possibility of an interaction between the pressure and the temperature,
both of which are required for the calculation of the velocity. If it were
possible to estimate these errors, it would require in practice a considerable
effort before a systematic measurement could be made inside a furnace.
The possible resultant errors can be as high as 10% for high velocities
(40 m/s) and can be more than 25% for velocities less than 4 m/s.

Chapter 9

Direction Sensitive Impact Tubes

The probes which we have described up till now have to be orientated in the flow direction in order to give a true value of velocity. Frequently the direction of the velocity is not known and its determination is important for the measurements (flames with swirl). The velocity is then characterized by two angles ϕ and δ and a scalar magnitude V, or by its three components in a system of reference axes associated with the probe and the installation. In addition to these three unknowns, there is also the static pressure which is often required. These four values can be measured with the direction sensitive impact tube.

9.1 Principle

We have seen that the distribution of pressure around an obstacle placed in a uniform flow depends upon the Reynolds number, as well as the magnitude and direction of the velocity. In the domain where most of our measurements are made (Reynolds number between 100 and 60 000, flow of type (c)), it has been established experimentally that the reduced velocity field V_{local}/V_0 around a sphere, placed in a uniform flow with velocity V_0, varies only slightly with Reynolds number. In other words, according to the Bernoulli theorem, the pressure, divided by the velocity squared, on the surface of a sphere is similar in form when the flow direction does not change. The pressure p_n, measured at a point n, is therefore given by the expression

$$p_n + \rho \frac{V_n^2}{2} = p_s + \rho \frac{V_0^2}{2}$$

$$p_n = p_s + \rho \frac{V_0^2}{2}\left[1 - \left(\frac{V_n}{V_0}\right)^2\right]$$

and, if

$$k_n = 1 - \left(\frac{V_n}{V_0}\right)^2$$

$$p_n = p_s + k_n \times \rho \frac{V_0^2}{2} \tag{9.1}$$

where V_n is the velocity at point n and k_n is a factor that only depends upon the velocity direction characterized by the angle θ_n between the axis and the pressure tapping n, or alternatively by two angles ϕ and δ in a system

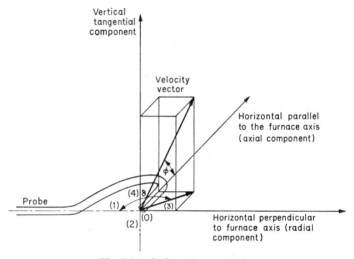

Fig. 9.1 Velocity reference angles

with axes fixed with respect to the installation (Fig. 9.1). When there are a sufficient number of pressure tappings, n, the unknowns V_0, p_s, ϕ and δ in the system of equations 9.1 can be calculated if the function $k_n(\phi, \delta)$ is known.

It is known that in the vicinity of the forward stagnation points the law is close to that corresponding to potential flow:

$$V_n = (\tfrac{3}{2} \sin \theta_n)V_0$$

and $$k_n = 1 - \tfrac{9}{4} \sin^2 \theta_n$$

However, there is a rapid deviation from this law, when θ_n increases, and the law becomes completely different in the vicinity of and downstream of the points of separation. In addition the condition of the surface of the sphere can influence the factor k_n; also the flow is never perfectly symmetrical around the probe and it is never possible to perfectly reproduce

a probe. Consequently the law $k_n(\phi, \delta)$ must be determined experimentally
for each instrument.

9.2 Five-hole probes

The method of measurement used at the Foundation is that first used by
Lee and Ash (17) for spheres with five holes, as shown in Fig. 9.2, but

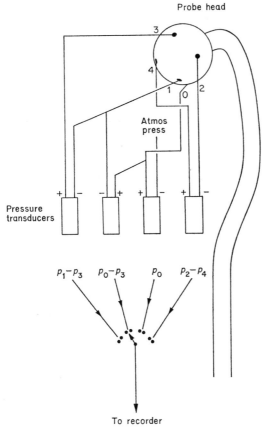

Fig. 9.2 Five-hole sphere

which is also applicable to tubes with a hemispherical end. The holes are
numbered 0 for the central hole and 1, 2, 3, 4 as shown in the figure. The
velocity direction is determined at first by eliminating the static pressure
by measuring pressure differentials

$$p_n - p_m = (k_n - k_m)\frac{\rho V_0^2}{2}$$

and subsequently the velocity magnitude is determined by the ratio of the pressure differentials:

$$\frac{p_1 - p_3}{p_0 - p_3} = \frac{k_1 - k_3}{k_0 - k_3} = f(\phi, \delta)$$

$$\frac{p_2 - p_4}{p_0 - p_3} = \frac{k_2 - k_4}{k_0 - k_3} = g(\phi, \delta)$$

The relation between the functions g and f and the angles ϕ and δ were established experimentally as shown in the calibration curve, Fig. 9.2, by varying the orientation of the probe placed in a constant velocity stream (18). The required angles ϕ and δ are determined from the pressure ratios and the calibration curve.

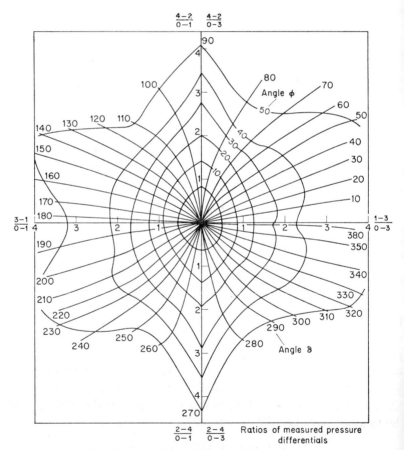

Fig. 9.3 Calibration of five-hole probe; measurement of angles

The equations can be used when the stagnation point is on the hemisphere 2–0–4–1 ($p_1 - p_3 > 0$) while in the other hemispheres the functions g and f become infinite for $p_0 = p_3$. We then use for $p_1 - p_3 < 0$, the functions

$$\frac{p_1 - p_3}{p_0 - p_1} = f'(\phi, \delta) \quad \text{and} \quad \frac{p_2 - p_4}{p_0 - p_1} = g'(\phi, \delta)$$

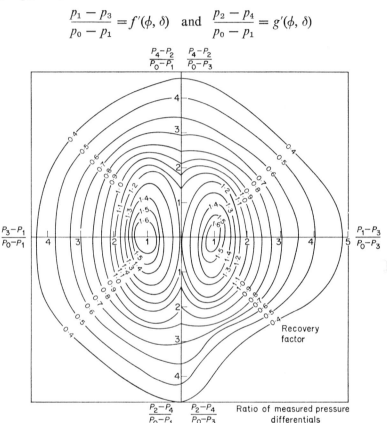

Fig. 9.4 Calibration of five-hole probe; measurement of velocity magnitude

Because of the instabilities of the flow in the vicinity of the points of separation, and also because of the influence of the probe support, the precision of the measurement diminishes as ϕ increases (calibration curves are generally only given up to $\phi = 60°$ for this reason). With the probes used at the Foundation, the possible error in the angles ϕ and δ is of the order of one degree (**10**).

The velocity magnitude is then calculated from the pressure differential:

$$p_0 - p_3 = (k_0 - k_3)\frac{\rho V_0^2}{2}$$

the coefficient $(k_0 - k_3)$ having been measured experimentally during the calibration of the probe, for each value of the angles ϕ and δ, *i.e.* the functions g and f (Fig. 9.4). Finally the static pressure p_s is obtained from one of the pressure measurements and the atmospheric pressure:

$$p_{atm} - p_s = p_{atm} - p_0 + k_0 \frac{\rho V_0^2}{2}$$

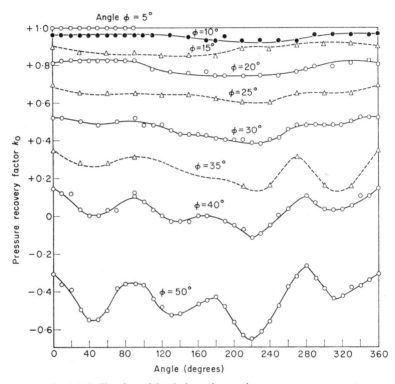

Fig. 9.5 Calibration of five-hole probe; static pressure measurement

k_0, like the other factors, is measured experimentally for each possible combination of the angles ϕ and δ, Fig. 9.5.

This method has the disadvantage that it takes a considerable amount of time to calibrate the probes and to calculate the results. It is for this reason that a new method has recently been developed which is more simple and more rapid (**19**). The aim was to simplify the calibration and to allow the calculations to be made with an electronic computer using the five measured pressure differentials: $p_0 - p_3$, $p_1 - p_3$, $p_0 - p_4$, $p_2 - p_4$, $p_0 - p_{atm}$.

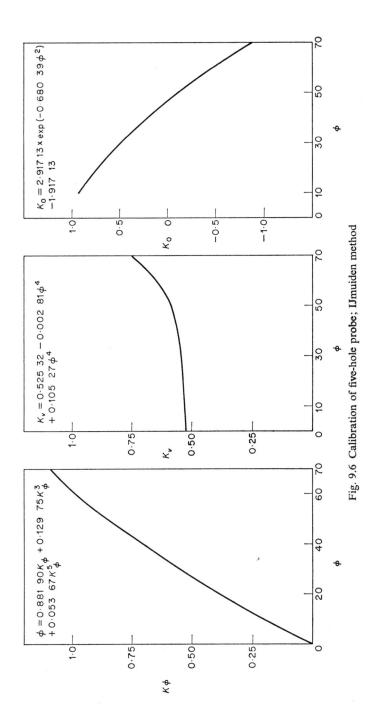

Fig. 9.6 Calibration of five-hole probe; IJmuiden method

In the first instance for a perfect sphere, the angle δ is given by the relation

$$\delta = \text{arc tg} \frac{-(p_2 - p_4)}{p_1 - p_3}$$

This equation is used after it was carefully verified in a wind tunnel that the proposed probe is sufficiently accurately manufactured so that the errors are not more than one degree. If this is found not to be the case, the instrument is not used. Subsequently the angle ϕ is calculated according to the relation:

$$\phi = A_1 K_\phi + A_3 K_\phi{}^3 + A_4 K_\phi{}^5 \qquad \text{(Fig. 9.6)}$$

or

$$K_\phi = \sqrt{1 - \frac{(p_0 - p_1) + (p_0 - p_2) + (p_0 - p_3) + (p_0 - p_4)}{2\sqrt{(p_0 - p_1)^2 + (p_0 - p_2)^2 + (p_0 - p_3)^2 + (p_0 - p_4)^2}}}$$

and A_1, A_3, A_4 are constants characteristic of the probe calculated according to the curve $\phi(K_\phi)$ obtained from the calibration for different values of ϕ and δ. Once again, the probe is only accepted when the experimental points are close to the mean curve chosen originally. The velocity is then calculated by the formula:

$$V = \sqrt{\frac{K_v}{\rho}} \times \sqrt{(p_0 - p_1)^2 + (p_0 - p_2)^2 + (p_0 - p_3)^2 + (p_0 - p_4)^2}$$

where $K_v = B_0 + B_2 \phi^2 + B_4 \phi^4$.

B_0, B_2, B_4 are constants measured during the calibration in a similar manner to A_1, A_3, A_4. The static pressure can be calculated according to the equation

$$p_{st} - p_{atm} = p_s - p_{atm} - k_0 \frac{\rho V_0^2}{2}$$

with

$$K_0 = C \times \exp(-D\phi^2) - E$$

C, D and E being constants corresponding to the mean curve $k_0(\phi)$ determined during calibration.

9.3 Orientatable probes—three hole probes

Another method to simplify the calibration and the calculation of the results is to orientate the probe until the pressures p_2 and p_4 are equal (Fig. 9.2). One of the angles is thus obtained directly, and the second angle is then found from the measurement of the pressure ratios $(p_2 - p_3)/(p_0 - p_2)$

and the value of this ratio obtained during a previous calibration. The velocity magnitude is then calculated from

$$V = \sqrt{\frac{K_v}{\rho} \times (p_0 - p_3)}$$

and the static pressure from

$$(p_{st} - p_{atm}) = (p_0 - p_{atm}) - k_0 \frac{\rho V^2}{2}$$

where the constants K_v and k_0 depend only on the ratio

$$\frac{p_1 - p_3}{p_0 - p_2}$$

It would also be possible to construct a probe which could be orientated in two perpendicular planes, giving directly the velocity direction and the static pressure, and then allowing the velocity magnitude to be calculated by the Pitot method. In practice, particularly for measurements in flames, such a probe would, on the one hand, be very difficult to construct, and on the other hand would require long measuring times in order to obtain equal pressures, particularly in turbulent flows.

Velocity measurements in two dimensional or axi-symmetric flows where one velocity component is zero, can be made with a three-hole probe.

9.4 Causes of error

The same causes of error as given previously for the Prandtl tube apply to the multi-directional probes while, in addition, errors arise due to inaccuracies in the calibration and changes which may occur in the characteristics of the probe head during service as a result of expansion and deposition of particles. Calibration of probes at IJmuiden are made with Reynolds numbers in the region of 3500, and it has been verified that the results are valid in the domain of $1000 < (Re) < 60\,000$. They are not valid for the very high Reynolds number region $((Re) > 200\,000)$ and neither for the very small Reynolds number range $((Re) < 100)$. Changes in turbulence intensity, however, do not seem to sensibly affect the measurements of velocity direction; calibrations made in the downstream region of a jet where the intensity of turbulence was 20%, gave the same result as those obtained in the calibration made in the potential core of the jet where the intensity of turbulence was only of the order of 1% or 2% (10).

When the 5-hole probes are placed in regions of high velocity gradient the stagnation point is displaced towards the region of highest velocity and errors then become dependent upon the dimension of the instrument **(20)**. It is for this reason that the probe heads have been reduced in size to $\phi = 4$ mm for cold isothermal work and for measurement in the furnace to $\phi = 12$ mm.

Chapter 10

Velocity Measuring Tubes Used at IJmuiden

A number of different types of velocity probes (Prandtl tubes, streamlined tubes, five-hole or three-hole direction sensitive probes) have been used at IJmuiden, according to the particular measuring conditions such as hot or cold models, pulverized coal or oil furnaces, see Table 10.1. For studies in cold models, uncooled probes of small dimensions and finely constructed are used. For measurements in the furnace the probes need to be more robust, cooled by water or constructed of refractory material and protected from condensation of water or oil vapour, and from the deposit of particles which can alter the characteristics of the probe and block the pressure hole (21)(22)(23).

These instruments are easily adaptable for use in most types of industrial furnace where the ambient conditions are similar to those in the IJmuiden experimental furnaces (Table 10.1).

10.1 Prandtl tubes

The standard dimensions of this probe have been given previously (page 184). It is not always possible to construct probes according to the standard dimensions because of the particular measuring conditions, and the tube constants must then be determined from a calibration in a wind tunnel (21).

It has been found that the constant $\dfrac{k.\Delta p}{\rho V^2/2}$ remains practically equal to unity when there are only minor modifications, which has been the case for most of the tubes described below.

Non-cooled tubes

These tubes are used essentially in models. Attempts have also been made, without success, to construct probes from refractory materials for use in furnaces.

Tubes for cold models. These tubes are of very small dimensions (minimum diameter used at IJmuiden is 1·2 mm, while diameters between 1·6 mm and

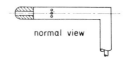

normal view

Use	No	Outer diameter		Material	Shape and details of head
Cold model	1	1·2 mm 1·8 mm 2·5 mm	Uncooled K = 1 ± 0·025 400 < Re < 700 000	Stainless steel hyperdermic needle	
Hot models	2	6 mm	Uncooled	Stainless steel soldered in the presence of argon	
COAL AND OIL FURNACE	3	12 mm 19 mm	Cooled (old model)	Ordinary steel and brass	
	4		Uncooled head (new model)	Ordinary steel and brass head of stainless steel	
	5		Uncooled	Alumina Al_2O_3, hot joints of refractory cements, rubber joints	
	6	6 mm	Cooled	Ordinary steel rubber joints	
	7		Normalisation		d 0·3d — 3d — 7 to 10 δ

Fig. 10.1 Prandtl tubes

2 mm are generally used), made of stainless steel hypodermic tubing with silver solder used in the probe head, see Fig. 10.1, No. 1. Calibration of such a tube carried out in the Reynolds number range between 400 and 700 000 has shown that the constant remains equal to 1 within ±0·025 (**21**).

Tubes for hot models. These tubes (Fig. 10.1, No. 2) with external diameters of 6 mm are also made with stainless steel, 18/8, and are constructed as the previous probes, but the soldering of the head is carried out under argon. When such a tube is placed in a hot gas flow, its temperature will depend upon convection (velocity and temperature of a gas), and especially on radiation (temperature of the model walls). It has been found that if the wall temperatures are less than 1000°C the tube can withstand a number of hours within the hot stream at a temperature of 1700°C and with velocities of several metres per second. It is important to limit the non-cooled length of the probe according to the measuring conditions, in order to avoid bending of the probe under its weight and that it should not be deformed due to the effect of the velocity. It is not normally necessary to go below 150 mm for a tube 6 mm in diameter.

Tubes for the furnace—stainless steel and alumina. The tubes described above have been used, because of their small obstruction, for measurements in a pulverized coal jet between the burner exit and the flame front where the particle density is very high (up to 1·5 grammes/N1) as well as the velocity (between 20 m/s and 40 m/s). The gas and particle temperature varies between 80°C and 1000°C, while that of the walls is in the vicinity of 1200°C. We have been able to establish that this tube can be used for a number of hours, except in close proximity to the flame front where the radiation as is high and the probe is subjected to chemical attack by tar and carbon, well as mechanical deformation under the effect of the compressed air, which it is necessary to blow periodically through the probe in order to clear the pressure holes. Its relative fragility, and particularly its proneness to blockage close to the burner, has made us abandon this tube in preference to the streamlined tube for measurement of velocity in this zone of the furnace.

In order to diminish the risk of blockage of the pressure holes, a probe has been constructed with a non-cooled head made of alumina which is heated to high temperatures by radiation so that carbon particles and oil droplets are burnt away and adhere less to the probe. The construction of this probe was found to be difficult because of the problems involved in forming a hemisphere at the head and also the necessity to obtain a good gas-tightness for the pressure lines at the junction with the water-cooled support. It was also found that the refractory tubes were easily broken because of their fragility to mechanical or thermal shock and so this tube is now no longer used at IJmuiden.

Water cooled tubes

These are the tubes that are normally used in furnaces. The standard tube has a diameter of 19 mm with an elbow support so that measurements are made along the axis of the probe in the same way as the temperature measurements (Fig. 10.1, No. 5). The total pressure hole has a diameter considerably smaller than that indicated by the standards, on the one hand in order to improve the cooling of the head of the tube and on the other hand to reduce the risk of blockage. This arrangement does not affect the measurement, provided that the tube is well orientated along the velocity direction, and that the velocity gradient is not too large, which generally is the case. Since cooling leads to condensation and deposit of particles, the entirely cooled tube is no longer used. It has been found preferable to use a tube with a stainless steel hemispherical head which is cooled over a length of 25 mm or 35 mm so that the temperature is raised to a value sufficient for the particles to burn. Cooling is effected by circulation of water, even when the probes are inserted in flames at the highest temperatures encountered in our furnaces. The standard version is very robust and is perfectly suitable for measurements in the flame or at the end of the furnace. Careful exploration of the jets in the vicinity of the burner, in particular in a pulverized fuel jet, require smaller probes in order to limit the perturbation at the measuring point and to avoid modification to the equilibrium of the furnace (such as breaking up of the pulverized fuel jet, displacement of the flame front, and the furnace equilibrium).

Prandtl tubes with an external diameter of 12 mm have been made sufficiently robust to replace the standard tube. On the other hand it appears to be difficult to make much further progress with miniaturization for industrial usage. This has been done by Pengelly (**21**) who has succeeded in constructing a tube with an external diameter of 6 mm in stainless steel (Fig. 10.1, No. 6). In this case the response time, because of the reduced dimension of the pressure holes, is no longer negligible. With an A.C.B. pressure transducer, with a small dead volume, it is of the order of several tenths of a second.

10.2 Special tubes

For certain velocity measurements, the Prandtl tube is not convenient because of its disturbance, its sensibility to blockage, its directional properties, etc. It is then necessary to construct an instrument which is better adapted to the particular experimental conditions. We give here a few examples.

Flat tube type BISRA

The tube shown in Fig. 10.2, No. 8, has the advantage of having a constant $k = \rho V^2/2\Delta p$, practically double that of the Prandtl tube.

Unfortunately its form causes perturbation in the flow and leads to blockage of the upstream pressure hole. It also is less accurate than the Prandtl tube and the constant k, which has to be established by calibration. is not constant at low Reynolds numbers.

Use	No	Size	Type	Material	
Hot furnace	8	10x70 x 150 mm	Flat tube BISRA type cooled $K \approx 1.80$		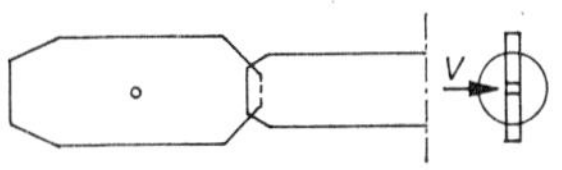
Velocity at hearth	9	14x 24 x 200 mm	Flow levels cooled $K = 1.35$	Ordinary steel	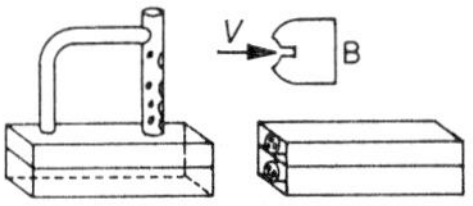
Hot furnace and coal jet	10	12 x18 x 150 mm	Cooled profile tube $K = 1.35$ $K = 0.35$	Ordinary steel	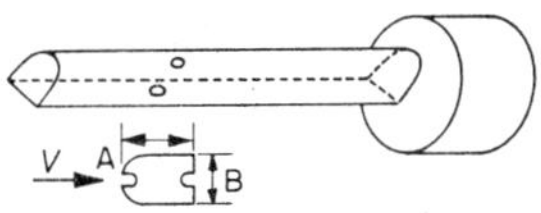
Static pressure in cold models	11	$\phi =12mm$ $l = 95mm$	S.E.R. disc uncooled	Ordinary steel	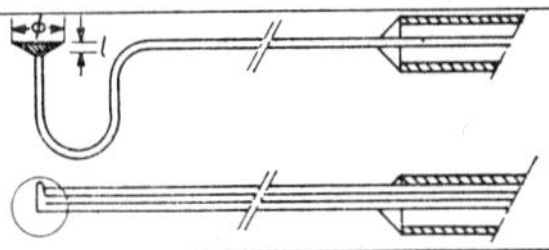
Cold models	12	ϕ head=? ϕ tube=?	5– hole impact tube uncooled	Ordinary steel	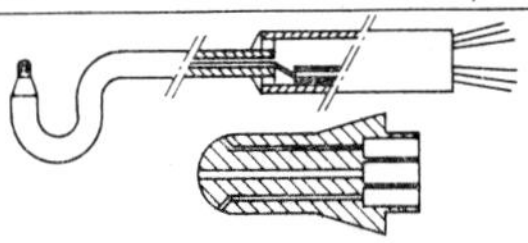
Hot furnaces	13	12 19mm	5– hole impact tube cooled with uncooled head	Ordinary steel stainless steel for the uncooled head	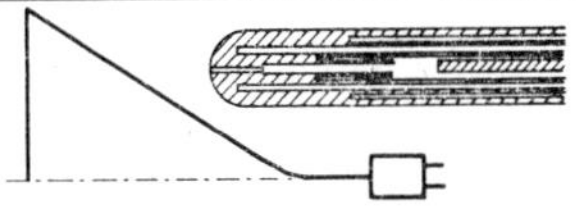
Cold models	14	ϕ of 20 and 50 mm	5–hole spherical impact tube cooled or uncooled	Copper or ordinary steel or stainless steel	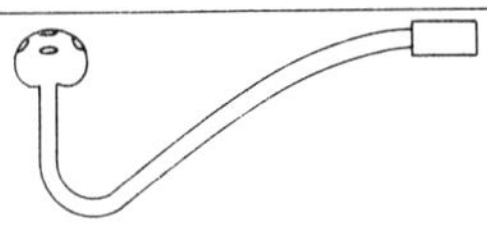

Fig. 10.2 Various tubes

Multiple tubes

In heat transfer experiments with impinging flames and local injection o oxygen, it has been necessary to measure the velocity profile in the vicinity of the furnace hearth. For this a probe was required to make measurements at four different heights, that could be introduced with little disturbance through furnace windows 25 mm wide and capable of withstanding very high temperatures (2000°C). An assembly of four water-cooled Prandtl tubes was not technically feasible and a measurement with four probes, each one supporting only one tube would have taken so much time that the number of measurements would have had to be reduced. The probe is shown in Fig. 10.2, No. 9. The head of the probe consists of a streamlined water-cooled tube with eight pressure holes (four on the upstream face and four on the downstream face). The constant of this tube is

$$\frac{\rho V^2}{2\Delta p} \simeq 1\cdot35 \ (\text{within} \ \pm0\cdot05)$$

and this constant is applicable for the four measuring heights (**3**). This tube was found to be only slightly sensible to the velocity orientation, which is particularly favourable in conditions when the flame is inclined. A system of electromagnetic valves fitted to the pressure lines allows the following operations:

(i)　blowing of compressed air simultaneously through the 8 pressure holes when blockage occurs;

(ii)　protection of the pressure transducer during blowing;

(iii)　successive connection of the four measuring points to a single pressure transducer;

(iv)　taking of a zero reading of transducer before and after each measurement.

This assembly which is very robust gives a larger response than a single Prandtl tube but, however, with a lower degree of accuracy. A tube of this type called "by the blade of a sword" (**9**), is used in industry to make measurements in conduits when the opening for introducing the probe is only of small dimensions. It has a constant k in the vicinity of $1\cdot95$.

Streamlined tube for velocity measurement in a dust-laden jet

As a consequence of the almost instantaneous blockage of the total pressure hole, it is not possible to use a Prandtl tube for the velocity measurement in the initial region of a pulverized coal jet (page 194). This problem can be resolved by measuring the differential pressure between the static pressure and the depression in the wake of an obstacle where the pressure hole is not blocked. A better sensitivity is obtained by separately measuring the static pressure p_{st} (with a short streamlined tube, of the same form as described previously, and using the atmospheric pressure as reference):

$$(p_{st} - p_{at}) + (p_{at} - p_A) = p_{st} - p_A = k'\frac{\rho V^2}{2}$$

The tube used has a width of 12 mm (Fig. 10.2, No. 10) and its constant k' is in the vicinity of 0·35 ($k' = k - 1$). The determination of this constant needs to be precise in order to avoid significant errors. By this means, taking into account the mean density of the mixture (gas + particles), the velocities measured at the burner exit were found to be in agreement with the velocities calculated from the exit flow rates.

10.3 Anemoclinometer

Five-hole tube

The tubes used at IJmuiden have hemispherical heads. Each new tube has to be calibrated before use because it has not been possible to assure sufficient reproducibility of the probes during construction. Calculation of the results previously carried out with the aid of calibration curves are now carried out by electronic computer.

(1) *Tubes for the cold model*: As shown in Fig. 10.2, No. 12, the tubes are constructed with a hemispherical head in copper and a diameter of 4 mm fixed to a support 6 mm in diameter.

(2) *Tubes for the furnace*: These are constructed like the standard Prandtl tube with the head 19 mm in diameter in stainless steel, non-cooled over 25 mm or 35 mm (Fig. 10.2, No. 13) in order to reduce the risk of blockage. The hemispherical head in which the five holes are drilled must be constructed with care and should have a perfectly polished surface. In the furnace it is obvious that deposits of particles or droplets of oil can occur, and modify the local velocity conditions, and therefore the pressure, thereby causing a serious error in the measurement. Successive heating and cooling of the head does not appear to have a significant effect upon the calibration. Calibration curves made before and after a series of trials were found to coincide within ±3°. It is always possible to periodically verify in a wind tunnel a few points on the calibration curves. Some of the tubes have supports which allow the tubes to be orientated in the plane of the velocity.

Three-hole tube

This is a cylindrical tube of 12 mm diameter, completely cooled, constructed in stainless steel, in which the two lateral holes make an angle of 45° with the central hole.

Pressure lines

These allow the transmission of pressure from the measuring point to the pressure transducer. They must therefore be perfectly gas-tight and should be sufficiently rigid so as not to transmit external pressures which may be applied to them. One may use metal tubes (copper or stainless steel) plastic or thick rubber. The lines can have an influence on the measurement either by acting as a resonance vessel and amplifying certain wave frequencies, or by delaying the signal if the volume is not negligible (24). The establishment of pressure equilibrium requires the displacement of a certain volume of fluid, which requires a time proportional to the dead volume in the conduits and the manometer as well as pressure loss in the conduits. The delay can be particularly sensitive when the conduit is constricted or partially obstructed at a point. The response time for the pressure measurement is often determined by lines and the dead volume of the transducer. For the A.C.B. transducers which have a high frequency and a small dead volume, lines of 1 mm to $1\frac{1}{2}$ mm internal diameter are most suitable. For correct measurements of a fluctuating pressure differential the response time of the two pressure lines have to be the same.

10.4 Measurement of pressure with pressure transducers

For a considerable number of years (21) the traditional manometers with liquid displacement have been replaced at IJmuiden by manometers with flexible membranes and electric transmitters based on the variation of mutual induction as supplied by A.C.B. (Ateliers de Construction de Bagneux). This system has the following advantages:

(i) very high sensitivity (measurement of 1/100th of a millimetre of water column);

(ii) small dead volumes, therefore instantaneous response;

(iii) electric signal which allows measurements to be taken at some distance on a potentiometer and which allows recording;

(iv) a number of ranges of sensitivity for one transducer;

(v) simple and rapid installation with little disturbance.

The measuring system consists of a flexible membrane with the pressure differential applied to the faces of the membrane, a variable mutual induction electric transmitter supplied with 1000 hertz, which translates the displacement of the membrane to an electric signal, a demodulator which rectifies the 1000 hertz current and filters parasitic currents, an attenuator, and finally an electric receiver (Fig. 10.3) (**25**). We shall discuss

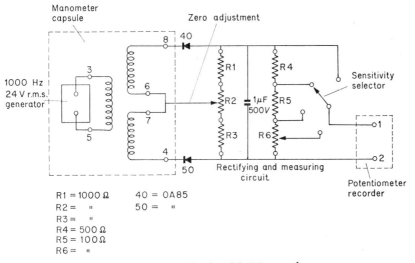

R1 = 1000 Ω   40 = 0A85
R2 =   "   50 =   "
R3 =   "
R4 = 500 Ω
R5 = 100 Ω
R6 =   "

Fig. 10.3 Electrical circuits of A.C.B. transducer

the characteristics and performance of the system as well as some elementary rules in order to obtain good measurements (**22**)(**26**).

Flexible membrane transducers

The displacement of the membrane should be a linear function of the applied differential pressure, and should not depend upon the ambient conditions: temperature, acceleration, level of pressure, noise, etc. Compensation for these conditions can be made but these reduce the simplicity of the instrument. The most classic example is that of temperature, generally effected not by the construction of membranes with a constant Young's modulus, but electronically by means of resistances or thermistors. A major inconvenience of this method is that the measuring components and the compensator components are not always necessarily at the same temperature, particularly during a period of transition: a sudden increase of the ambient temperature by 50°C is not fully compensated until after a wait of half an hour (with the standard transducers of

A.C.B. the temperature level has no influence on the measurement between 0°C–50°C, but it is necessary that the transducer and demodulator remain at a constant temperature). When the transducers have to be used under difficult experimental conditions, it is recommended that the manufacturer is advised so that any changes which are considered necessary and possible can be made (27).

The membranes constructed of different materials (bronze or beryllium, silver, nickel-chrome alloy, stainless steel, etc.) undergo constraints which are well distributed over all their surface and the displacements are always very small, so that the error due to hysteresis is in principle negligible (0·1 %). This can moreover be diminished by submitting the membrane to a few cycles of pressure with a slight overload. For industrial use it is particularly useful to have transducers that are able to support large overloads. For this purpose the movement of the membrane is limited by a fitted stop (for overloads up to a hundred times the nominal scale of measurement) or alternatively to use two membranes stuck together (for overloads up to one thousand). At IJmuiden the transducers used are those that give ±2, 5, 20, 50 and 100 mbar at full scale. A.C.B. however, have a more extensive range of pressures.

Measurement of displacement by variation of mutual induction: electrical transmitter

The transmitter is made up of two symmetrical transformers with each fixed magnetic circuit having an air gap. A single magnetic core displaced by the membrane modifies the two air gaps, by its movement, in the opposite direction. The primaries of the two transformers are connected in series, and at the terminals of the independent secondaries, two voltages are picked up which vary in the opposite sense to the air gaps. When the displacement is small, it can be shown that the potential difference of the secondaries is:

$$E_2 - E_1 = \omega i(M_2 - M_1)$$

with $M_2 - M_1$ proportional to $2x/e^2$

where ω is a frequency of the primary current,
   i is the amplitude of the primary current,
   e is the fixed air gap, and
   x is the variation of the air gap—a variation proportional to the displacement of the membrane, and linearly dependent on the measured pressure difference.

(1) *Stabilized supply*: The above formula shows the necessity of having a stabilized supply, and E and i should remain constant. An alternating current generator of 1000 Hz or 3000 Hz (10 000 Hz has also been used), under a regulated voltage of 22 volts, supplies 0·5 to 1 VA per transducer. This generator may be a rotating machine or an ensemble of electronic valves or transistors. The latter have the advantage of being more stable and are recommended for experiments of long duration. The choice of frequency is dependent on the sensitivity and response time required. The passing band is limited to one-fifth of the carrier frequency (limitations may also be imposed by the transducer and the pressure lines).

(2) *Demodulator*: The secondary exit voltages, E_1 and E_2 are rectified and placed in opposition. A variable resistance allows the secondary circuits to be balanced before the measurement to obtain their zero. The temperature coefficient of the rectifier is compensated by that of copper electrical resistances. A by-pass filter is added in order to eliminate the residuals of the carrier frequency.

(3) *Receiver*: The receiver should have an impedance of about 1000 ohms capable of measuring potential differences of 250 mV. Adaptors can be fitted which allow the use of potentiometric recorders, cathode ray oscilloscopes, transmitters, etc. At IJmuiden, a potentiometric recorder is used, placed at the terminals of an attenuator with four sensitivities.

Calibration of the pressure measuring network

Calibration is carried out by submitting the transducer to a series of known pressure differentials, for example, by means of an Askania screw

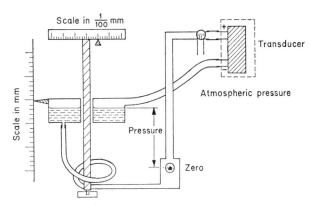

Fig. 10.4 Diagram of Askania minimeter

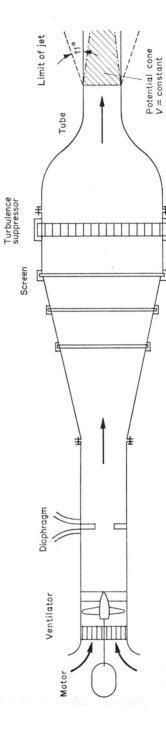

Fig. 10.5 Diagram of wind tunnel

minimeter (Fig. 10.4). It can be verified, by this means, that the characteristic calibration of millivolts as a function of pressure is linear. The two millivolt transducer, without attenuation, gives more than ten millivolts per millimetre of water column.

10.5 Calibration wind tunnel

For the calibration of velocity measuring instruments an air flow is required which is stable with time and sufficiently large not only to contain the probe to be calibrated, but also sufficiently large so that this probe will not disturb the flow. The flow may be confined in a wind tunnel, or allowed to discharge to the atmosphere (Fig. 10.5). In the latter case measurements must be made within the potential core of the jet. Uniform exit conditions can be obtained by fitting a convergent nozzle as shown in Fig. 10.5. Further examples of wind tunnels can be found in references (5)(28)(29).

Bibliography

1. PENGELLY, A. E. S. New equipment for flame and furnace research. *Journ. of the Inst. of Fuel*, **35**, p. 210, May, 1962.

2. KISSEL, R. R. (a) *Appareils de mesure actuellement utilisés ou en cours de mise au point pour l'étude des flammes à IJmuiden*. I.F.R.F. IJmuiden, January, 1960. Doc. nr. F 72/a/4. (b) *La Fondation de Recherches Internationales sur les Flammes son organisation. Description de la station d'IJmuiden et son appareillage de mesures.* Silicates Industriels, pp. 451–467.

3. CHEDAILIE, J. and BRAUD, Y. Equipements, méthodes et instrument nouveaux mis en service à la Station d'IJmuiden en 1964 et 1965 (*6ème Journée d'Etude sur les Flammes*, Paris Comité français, 6 Rue Cognac Jay, Paris, Vlle), November, 1965.

4. STOLL, H. W. The Pitot venturi flow element. *Trans. of the A.S.M.E.*, Vol. **73**, pp. 963–969, October, 1954.

5. PRANDTL, L. *Essentials of Fluid Dynamics*. Blackie and Son Ltd., London, 1954.

6. SCHLICHTING, H. *Boundary Layer Theory*. McGraw-Hill, 1968.

7. FOLSON, R. G. Reviews of the Pitot tubes. *Trans. of the A.S.M.E.* (with 129 references), Vol. **78**, pp. 1447–1460, October, 1956.

8. HUBBARD, C. V. Investigation of errors of Pitot tubes. *Trans. of the A.S.M.E.*, Vol. **61**, pp. 447–492, August, 1939.

9. BURTON, J. *Pratique de la mesure et du contrôle dans l'industrie*. Tôme 1. *Pressions-niveaux-débits*. Ed. Dunod, Paris, 1958.

10. BEÉR, J. M., CHIGIER, N. A., KOOPMANS, G. and LEE, K. B. Measuring instruments for the study of flames at IJmuiden. I.F.R.F., IJmuiden, Doc. nr. F 72/a/9, May, 1965.

11. BARAT, M. *Difficultés relatives aux déterminations des champs de pression et de vitesse dans les écoulements turbulents*. Eurovent, Paris, 1965.

12. HURD, C. W., CHESKU, K. P. and SHAPIRO, A. A. Influence of viscous effects on impact tubes. *Journal of Applied Mechanics*, pp. 253–256, June, 1953.

13. MACMILLAN, F. A. *Viscous effects on Pitot tubes at low speeds*. Aeronautical Research Council, Fluid motion Sub-committee, Report F.M. 2081 (8 pages), 1954.

14. GOLDSTEIN, S. A note on the measurement of total heat and static pressure in a turbulent stream. *Proceedings of the Royal Society of London*, 1936, Series A, Vol. **155**, pp. 570–575.

15. HEMSATH, K. H. and FABER, A. *Computer programme: Formulae used for the computation of data obtained from the trials C-11 and C-12.* I.F.R.F. IJmuiden, Doc. nr. G 00/a/2, March, 1966.

16. LEUCKEL, W. *Measurement of velocity in dust laden flow.* I.F.R.F. IJmuiden, Doc. nr. G 02/a/11, November, 1965.

17. LEE, J. C. and ASH, J. E. A three dimensional spherical Pitot probe. *Trans. of the A.S.M.E.*, Vol. **78**, pp. 603–608, April, 1956.

18. ROBERTSON, A. D. *The calibration of a spherical gas velocity probe.* United Steel Companies Ltd., Rotherham, England, F and Fr 4678/1/62 and I.F.R.F. IJmuiden, Doc. nr. F 72/a/1, November, 1962.

19. LEUCKEL, W. *A simple method to plot calibration coefficients of a hemispherical five-hole velocity probe.* I.F.R.F. IJmuiden. Doc. nr. F 72/a/12.

20. CHESTERS, A. K. *The influence of velocity gradients on five-hole Pitot measurements.* I.F.R.F. IJmuiden, Doc. nr. G 02/a/10, November, 1965.

21. PENGELLY, A. E. S. Apparatus for the measurements of gas velocity in furnaces and models. *Journal of Scientific Instruments*, Vol. **37**, September, 1960.

22. MAYER, G. and PENGELLY, A. E. S. *Improved Pitot tube for use in experimental furnaces.* I.F.R.F. IJmuiden, Doc. nr. F 72/a/2, January, 1957.

23. RIVIERE, M. J. M. *Mesure des vitesses dans les flammes.* I.F.R.F. IJmuiden, Doc. nr. F 31/a/4, November, 1954.

24. WATTS, G. P. *The response of pressure transmission lines.* Instrument Society of America. 20th Annual Conference and Exhibition, Los Angeles, 4–7 October, 1965.

25. SCHLUMBERGER, A. C. B. *Materiel de mesure à variation de mutuelle induction.* (Ateliers de Construction de Bagneux), 57 Rue de Paris, Bagneux (Seine), France.

26. GONDET, P. *Measuring to "one part in a thousand".* Schlumberger instrument exhibition, Lyon, A. C. B. Schlumberger (Ateliers de Construction de Bagneux), 57 Rue de Paris, Bagneux (Seine), France, October, 1964.

27. BABAUD, M. J. *Adaptation de certains capteurs de mesure de pression à des conditions d'emploi très sévères.* A. C. B. Schlumberger (Ateliers de Construction de Bagneux), 57 Rue de Paris, Bagneux (Seine), France.

28. LE MARECHAL, T. Une soufflerie chaude (500°C) à veine libre pour étalonnages. *Bulletin du Centre de Recherches d'Essais de Chatou, France*, No. 9, pp. 43–50, October, 1964.

29. MESRINE, R. Soufflerie à gaz très chauds. *Bulletin du Centre de Recherches et d'Essais de Chatou, France*, No. 16, pp. 39–55, June, 1966.

30. OWER, E. and JOHANSEN, F. C. *The Design of Pitot-static Tubes.* Great Britain, ARC, R and M 981, 12 pp. + 11 pp. of figs., 1925.

31. HOMANN, F. Der Einfluss grosser Zähigkeit bei Strömung um den Zylinder und um die Kugel. *Zeitschr. für angen Math. und Mech.*, Vol. **16**, pp. 153–164, 1936. Translation published: NACA TM 1334, 29 pp., 1952.

32. CHAMBRE, P. L. and SMITH, H. R. The Impact Tube in a Viscous Compressible Gas. Univ. of Calif., Berkeley. *Heat Transfer and Fluid Mechanics Inst. Publ.*, pp. 271–278, June, 1949.

33. IPSEN, D. C. Viscosity Correction to Impact Pressure on Prolate Spheroid. Univ. of Calif., Berkeley, *Inst. of Engineering Res. Rep.*, He. 15089, 15 + 3 pp., 1952.

Index